A
QUARTER
MILLION
STEPS

CREATIVITY, IMAGINATION, &
LEADING TRANSFORMATIVE CHANGE

A QUARTER MILLION STEPS

Enjoy the Journey!

ANTHONY PAUSTIAN, PH.D.

Foreword by Apollo 15 Astronaut Alfred Worden

Book Press™ publishing

www.bookpresspublishing.com

Published in Des Moines, Iowa, by:

BookPress Publishing
P.O. Box 71532 • Des Moines, IA 50325
www.BookPressPublishing.com

Publisher's Cataloging-in-Publication Data

Names: Paustian, Anthony D., 1964-, author.
Title: A Quarter Million Steps : creativity , imagination , and leading transformative change / Anthony Paustian, Ph.D.
Description: Includes bibliographical references. | Des Moines, IA: BookPress Publishing, 2017.
Identifiers: ISBN 978-0-9964428-9-3 | LCCN 2016917192
Subjects: LCSH Leadership. | Organizational change, | Transformational leadership. | Creative ability in business--Management. | Teams in the workplace. | Influence (Psychology) | BISAC BUSINESS & ECONOMICS / Leadership.
Classification: LCC HD57.7 P38 2017 | DDC 658.4092--dc23

First Edition
Printed in the United States of America
10 9 8 7 6 5 4 3 2 1

This book is dedicated to the over 400,000 men and women who, because of their imagination, creativity, dedication, focus, and exceptional work ethic, inspired not only me but my entire generation.

CONTENTS

Foreword

by Alfred (Al) Worden
Apollo 15 Command Module Pilot

Long before becoming an astronaut and taking the quarter-million mile journey to the Moon, I grew up on a small farm in Jackson, Michigan. One night, while at my grandfather's farm in northern Michigan, I awoke to the intense thunder of a terrific storm, rattling

Apollo 15 Astronaut Alfred Worden inside the Command Module Endeavour.
(Image courtesy of NASA)

the windows and raining buckets. I was very young and therefore concerned for the animals in the fields and barn. I went to my grandfather and asked him what we should do. He told me not to worry; he had already moved all the livestock into the barn and closed all the windows after completing the chores. He taught me to look out for those who worked for me, and that if I took care of them, they would take care of me. It was a great lesson in responsibility, and I've never forgotten his philosophy.

Like my grandfather, Dr. Paustian teaches the need for sound thinking in every aspect of our lives and what it takes to accomplish something great like we did during the Apollo Program. As I read this book, I thought about what it takes to accomplish challenging goals and the quarter-million steps I took to achieve mine. No matter who you are or where you come from, this book will teach you what many of us take a lifetime to learn. Most importantly, it teaches the principles of leadership and taking personal responsibility for the well-being of others, just like my grandfather took care of those who worked for him.

As a young adult, I attended the U.S. Military Academy at West Point and then joined the Air Force. I had never flown much before, but I thought it would be an exciting and rewarding career. Flying turned out to be something I could really relate to. During my training, I realized I was good at all-weather flying and was ultimately assigned to an all-weather squadron in Washington, D.C. However, when I arrived, I found we didn't have enough aircraft in commission to fly a full schedule. It was a disappointment to say the least.

Instead of just hanging out in the pilots' lounge, I wandered around the repair shop for electronic maintenance. It was a mess, and I began to think that part of the problem was its deplorable condition. The squadron commander saw my interest and asked me to manage the shop, as he had no confidence in the currently assigned Armament Officer. I believed I could do better, so I said yes. It was a matter of promoting quality maintenance while creating ways to accomplish it. It was an example of expanded "in the box" thinking, one of the many skills *A Quarter-Million Steps* will teach you.

The shop was required to maintain the aircraft as prescribed by the manuals, but it was not being done well. Instead of berating and blaming the technicians, I decided to do something entirely different. I focused on changing the work environment. I believed that if the shop was as clean as an operating room, then the work habits and performance of the technicians would naturally improve. Funding was also an issue, so I convinced the companies who supplied our aircraft and avionics that I needed their help. My plan worked. We ended up with a squeaky-clean shop with good lighting and polished floors. It was amazing to see the change in the work ethic and performance of the technicians. Our in-commission rate went up to nearly 80%, and flying became a joy.

Because of the creative approach we took to problem solving at the squadron, I was asked to transfer to headquarters and take my message to all the fighter squadrons in the Air Defense Command. Instead, I requested and received an appointment to go back to college and get my master's degree.

My philosophy has always been to do the very best you can at whatever you set your mind to, something Dr. Paustian also teaches in *A Quarter-Million Steps*. It just so happened that while I was working on my degree, I became friendly with two instructors from the Air Force Test Pilot School at Edwards Air Force Base in California. They convinced me to apply for the test pilot school; I not only had the education, but while going to college, I was also logging a lot of flying time as the Operations Officer. Since I was driven to be the best pilot in the Air Force, I applied and was accepted.

I was on a path to become a successful test pilot when NASA called for applicants to the space program. Because I felt I was too old for the program, I never thought too much about it until other doors were opened. After working so hard to become the best test pilot, my work ethic was rewarded in a way I didn't expect. I was accepted by NASA and went on to fly in space on Apollo 15, one of only a handful of flights to the Moon and the pinnacle of my career.

Now at the age of 84, I am reviewing my life and the many events that provided direction and purpose to my adventure. The same lessons I learned through hard work and decades of experience are the lessons Dr. Paustian outlines in this important book. You will find great examples and suggestions to help you live a full and successful life, particularly in the area of "imagineering," something I worked hard to achieve. You'll find techniques and strategies to harness your creativity and imagination, as well as strategies for effective leadership and motivating a team to achieve HARD goals.

Dr. Paustian and I both believe that if you follow your dreams you can make a difference in the world, and this book will show you how to do it. Work hard at excellence, and you will be surprised at the doors that will open for you.

Why a Quarter-Million Steps?

Whether it's building the Panama Canal, inventing new technologies in Silicon Valley, designing massive skyscrapers for New York City, developing new wind energy in the Plains, creating larger-than-life movies in Hollywood, growing food more efficiently for the masses in Iowa, or advancing medicine at Johns Hopkins Hospital or the Mayo Clinic, the people of this country have continually, throughout history, realized incredible feats of imagination.

However, in my opinion, nothing has been more momentous than when more than 400,000 people worked together to put a man on the surface of the Moon during the 1960s, a decade of challenging social unrest. The country during that period in history coped with a very unpopular war in Vietnam, a Cold War with the Soviet Union, the fear of nuclear annihilation, the assassinations of both a president and the leader of the Civil Rights Movement, sweeping national protests, violent riots, a stagnating economy, sensationalized serial killers, and a growing drug culture. Despite it all, the National Aeronautics and Space Administration (NASA) through the Apollo Moon Program was able to orchestrate a massive creative effort to achieve a feat that would be difficult, if not impossible, to replicate today.

The Apollo Program was not socially popular during its time, and a majority of the country didn't want to fund it or the space program in general. But, while Apollo was making global headlines when Neil Armstrong made the first boot impression on the lunar surface, it was also driving science and creating thousands of new ideas that ultimately resulted in a significant positive economic impact. Much of this *imagineering* resulted in new commercial products and the foundation of entire industries such as the microchip and the creation of Intel, microwaves, smoke detectors, cordless power tools, cool suits used by firefighters and race car drivers, kidney dialysis and CAT/MRI scanners used in healthcare, solar panels, quartz timekeeping, polarized sunglasses, water purification, freeze-dried foods, and the list goes on. Economic studies on Apollo show a societal return on investment of up to $14 for every $1 spent, making the returns on most other forms of investment pale in comparison.[1,2]

The Apollo Program taught us how important it is to keep the lines of communication open with all stakeholders and to be strategic about team communication. It showed how effective planning is the single most important step to success, but that we also shouldn't be afraid to modify that plan when necessary. It revealed how risk should be acknowledged and considered but not serve as a deterrent. It demonstrated methods to successfully delegate tasks and challenge people in an environment filled with unknowns. Since no one had ever done this before, tasks were frequently delegated to people without any experience who were left to figure it out on their own. Everyone took learned lessons to heart and demonstrated how to celebrate success as a unified team. In the words of Dr. Christopher Kraft, Director of Flight Operations for Apollo,

"We said to ourselves that we have now done everything we know how to do. We feel comfortable with all of the unknowns that we went into this program with. We don't know what else to do to make this thing risk-free, so it's time to go." [3]

The Apollo Program briefly united countries around the globe and made television king as an estimated 530 million people watched together on July 20, 1969, when Neil Armstrong took his "giant leap for mankind." It inspired, and it confirmed that if you dream it, arouse it, plan it, and finance it, almost *anything* can be accomplished.[4]

Ticker-tape parade celebrating the Apollo 11 mission and the first successful Moon landing.
(Image courtesy of Wikipedia Commons)

As time passes, the economic, political, business, and social impact of Apollo diminishes. For this reason, I chose to use the Apollo Program as a repeated metaphor throughout this book to discuss a variety of topics including **creativity**, **innovation**, **imagination**, **perception**, **transformative change**, **focus**, **teamwork**, **work ethic**, and **leadership**. Specific examples from the Apollo Program, including details from personal conversations with the astronauts who were personally part of it, are used throughout to help drive home many of the points and concepts that are now very relevant today.

The quarter-million mile journey to the Moon was the culmination of an enormous amount of planning, effort, failure, and success. What began as one leader's vision resulted in one of mankind's greatest leaps accomplished through a quarter-million steps taken one at a time.

Beware the Purple People Eaters

Had Kennedy lost his mind? He'd only been President for a few months. The country was engaged in a full-on Cold War with the Soviet Union, forcing many Americans to build backyard fallout shelters. The Civil Rights Movement was working hard to end racial segregation and discrimination through widespread sit-ins and civil resistance. The failed invasion of Cuba at the Bay of Pigs a month earlier was an unmitigated disaster fresh on everyone's mind. But, there he was, May 25, 1961, standing before a joint session of Congress, boldly proclaiming something many thought to be absolutely absurd:

> *"...I believe that this nation should commit itself to achieving the goal, before this decade is out, of landing a man on the Moon and returning him safely to the earth. No single space project in this period will be more impressive to mankind, or more important for the long-range exploration of space; and none will be so difficult or expensive to accomplish. We propose to accelerate the development of the appropriate lunar*

spacecraft. We propose to develop alternate liquid and solid fuel boosters, much larger than any now being developed, until certain which is superior...But in a very real sense, it will not be one man going to the Moon—if we make this judgment affirmatively, it will be an entire nation. For all of us must work to put him there." [1]

At the time of John F. Kennedy's speech, the United States had only put one man in space...for a grand total of 15 minutes and 28 seconds. In a cannonball-like shot, Alan Shepard had barely enough time to look down at the Earth by the time he was back standing on it.

Was President Kennedy delusional? That speech was completely illogical and had no basis in reality considering the lack of available technology, the proposed timeframe, and the required resources to make it happen. We hadn't even put a man in orbit around Earth yet, let alone anything resembling the quarter-million mile trek to the Moon.

Fast-forward nine years to July 20, 1969. I remember quite vividly sitting in front of my grandmother's large black and white television watching as both Neil Armstrong and Edwin "Buzz" Aldrin became the first men to step foot on another celestial body. What began as a monsterous, ambitious dream by one became reality for all. The phrase "We put a man on the Moon" is now the benchmark for how a shift in conventional thinking can cause almost *anything* to happen one tiny step at a time, or a quarter-million of them in this case.

Stay in Your Box

For years now, people have been using the phrase, "Think outside the box." This metaphor has been promoted and reinforced endlessly to encourage us to think in different or unconventional ways. The phrase is popular because it espouses an important idea, one that brings with it the appeal of expression (it's catchy, easy to remember, and easy to picture—in short, a marketer's dream). But is it time to reexamine the notion? In my various roles, I've pondered whether it's time to stop and take a look at what was initially in the "box" before everyone decided to start looking outside of it.

My conclusion: some basic *in-the-box* principles seem to have gone by the wayside in our quest to think *outside*. These are the very principles that provide a deeply rooted foundation, a foundation that allows us to venture beyond the box and look at things differently if desired, yet enables us to return to the box (and our core values) when necessary. Unfortunately, however, these time-honored values can get lost during the pursuit of what's outside. I continually observe a growing number of people who seem to have little to no awareness of the importance or benefit these principles add to long-term decision-making and problem solving.

I don't mean to imply that I have lost faith in the future. But, like others before me, my perspective has changed. I remember as a younger person being given advice and guidance by those much wiser: parents, grandparents, mentors, and supervisors, not to mention the "authority" figures in pop culture at the time, such as ninja masters, the Fonz, or the great Oz himself.

I'm sure—consciously or not—I dismissed some of what others tried to impart because I was young and cocky at best, ignorant at worst. Yet, these wise people have "been there, done that." They have learned from their mistakes and helped others avoid making the same. Sometimes I believe getting older is nothing more than God saying, "I told you so. You should have listened." I am now becoming one of "those" people—someone who, because of life experience and the mistakes that come with it, feels compelled to help others avoid the same mistake-laden path.

I fully understand that much of my new awareness is borne from one person's view based on over a half-century of anecdotal evidence. But, the need to reexamine cultural practices and beliefs (both past and present) has never been greater. It's tempting, I know, to simply assume nothing old works or is acceptable any longer and to dismiss traditional values as getting in the way of what's new. But, we do so at our peril.

We're always told to think outside the box, but perhaps the box itself is the key to success. NASA didn't immediately look to other countries to begin fulfilling Kennedy's vision. They built on what they had already accomplished here based on a prescribed set of values. Some of these values included a heavy reliance on U.S. private industry, universities, and research institutions; built-in redundancy for human safety; encouragement of internal competition to create a more precise and viable space exploration effort; and the "10% Rule" (a rule stating that ten percent of all NASA funding was to be spent on in-house expertise to ensure contractor quality). They also created a program management tenet requiring the most critical, intertwined factors—cost, schedule, and reliability—be managed as a group.[2]

Instead of looking outside the box, I believe we need to expand the size of the box we're already in through increased learning and higher levels of inspiration, creativity, innovation, and imagination.

Enter the Purple People Eaters

It's funny what you remember from early childhood. Along with sitting and watching the first Moon landing, I can also recall purchasing wax lips and Pixy Stix at the candy store as I walked home from school when I was six; how, at age seven, my second grade teachers would place me alone in a janitor's closet because I was too disruptive in class; and when I was "beaned" directly in the forehead by a baseball in third grade while running between two older kids in a game of pickle. I have often wondered what impact these memories have had, if any, on how I have turned out as an adult.

I realize who I am today is the result of my life experiences, but I can't help but wonder why some memories remain vivid, while others fade. One of my most vivid and clear memories occurred in first grade. My classmates and I were in the middle of coloring time. As I colored a person in my book with a purple crayon, my young, very attractive teacher (whom I also had a serious crush on) leaned over my shoulder, handed me the peach-colored crayon, and said, "Tony, there's no such thing as purple people. Could you please use a real color for their skin?" This comment not only trampled my "manly" self-esteem, it also stifled my creativity.

Even though this experience occurred long before the age of political correctness, my teacher was obviously more concerned my work was correct, rather than different. In short, she provided an example of what I now refer to as a *Purple People Eater*, something (usually a person, but also an idea, feeling, etc.) that eats away at the creativity (or Purple People) in others.

They're Everywhere!

The world is full of naysayers: downers with negative attitudes, rule enforcers, priority police, know-it-alls, people fearful of change, and those simply unable to shift their focus of reality. These are the people who continually bombard us with all of the reasons "why not;" those who serve as psychological roadblocks to inspiration, imagination, and creativity; and those who continually try to prevent others from doing something different, new, or outside the established boundaries—like those who told Kennedy that going to the Moon was impossible or impractical. These Purple People Eaters immediately shoot down ideas with phrases that begin with "You can't," or "It won't."

Make no mistake: I am not an idea zealot. I don't believe in change for the sake of change or in doing something just because it's new. I recognize not all things need to change, especially if no one has created a solution better than the one currently in place. However, until an idea has been fully vetted, this aspect of the equation is often unknown. As long as people are willing to let others immediately stifle new ideas, there's no way of knowing if the idea or solution is better than the current one. Sometimes, we just don't know until we try.

If Alexander Graham Bell had listened to the naysayers, the telephone might never have amounted to more than a "parlor toy." If Thomas Edison had listened to people bemoaning his lack of progress while designing the light bulb, I might still be writing this by candlelight. If Henry Ford had listened only to those who were unable to shift their focus, he would have tried to create a "better horse" instead of making the automobile affordable for the masses. If Albert Einstein hadn't received the support of Max Planck, recipient of the Nobel Prize in physics, $E=mc^2$ might never have been taken seriously by the physics establishment; Einstein could have finished his career as nothing more than a patent clerk. Perhaps I'm being a little extreme, as the outcomes of their efforts might have occurred in any case. However, these examples demonstrate that Purple People Eaters are everywhere—always have been, always will be. Avoid them.

Creativity: The New Currency

Imagination and creative thinking are on the decline. We can see this whether we're looking at the result of a study at the College of William & Mary detailing the downward trend of creative ability,[3] an IBM Corporation study identifying creativity as the number one "leadership competency" for the future,[4] or the rash of new books and articles suggesting that creativity has now become the new currency in business because of its short supply.

This, combined with my personal experience as both a leader and follower, only fuels my concern about the future of America's competitive ability in a rapidly changing and growing global economy. For more than two centuries, this country steadily led the world in creativity and innovation. Based on patent filings, a widely accepted measure of both, the United States was the dominant leader in global filings. However, from the early 2000s to the present, China's filings have been increasing at an annual rate exceeding 25 percent, while U.S. patent filings have been on the decline.[5] According to Thomson Reuters, China should become the world's leader in patent filings in the very near future, surpassing both Japan and the U.S.[6] Hence, at least in terms of this particular measurement, China will be considered the world's leader in creativity and innovation.

Why the U.S. decline? Some might say it's the countless hours that children and teens now spend in front of a television or computer watching videos or playing games instead of relying on their imagination. Others might say it's the lack of time kids have today, as they are shuffled from school to soccer to karate classes to flute practice to Cub Scouts. Perhaps it has more to do with how we define creativity, especially the concept of creativity as it relates to early childhood. Everyone is born creative, yet most people probably can't define exactly what creativity is or how it's accomplished. Many will frequently proclaim they lack it and may often view "creativity" as simply the ability to draw or paint, compose music, design houses, invent products, create movies, or demonstrate some other tangible outcome. Yet, true creativity encompasses much more than this limited view.

Let's Get Sticky!

When writing the book *Imagine!*, I defined creativity as "...simply a matter of seeing things that everyone else sees while making CONNECTIONS that no one else has made."[7] True discoveries seldom happen today by *finding* something new. Most often they are the result of "sticky thinking," which occurs when people connect or stick things together in new ways for different or improved outcomes. For example, elements such as oxygen, hydrogen, and carbon are some of nature's building blocks. The last element to be found existing naturally in the physical world was Francium in 1939, which was discovered by Marguerite Perey at the Curie Institute in Paris (and it was found to occur only in very trace amounts).[8] Any known element since then has been synthesized in a lab by sticking together two or more already known elements. This process of sticking or connecting things together in unique ways is the very basis of creativity.

Born in 1944 with a bone socket hip disorder called Calve-Perthes disease, young Fredrick Smith had to walk with the aid of braces and crutches for most of his childhood. But, through a high level of dedication and hard work, he was able to overcome the disease. In the early 1960s, Fredrick attended Yale University majoring in economics. For one of his classes, he wrote a paper detailing an idea he had after realizing that a future "automated society" required a completely different system of logistics. His professor didn't like the idea since, at the time, it wasn't economically feasible, but that didn't stop him from thinking about its future possibilities.[9]

After graduation, Fredrick went on to serve two tours in Vietnam as a platoon leader and narrowly survived a Viet Cong ambush. Upon returning from war, he wanted "to do something productive after blowing so many things up." Fredrick took an inheritance from his father, raised an additional $91 million in venture capital, and used the idea from his paper at Yale to create what is today known as FedEx. Fred Smith's story is arguably one of the greatest entrepreneurial successes of the last century. His net worth is in the billions, and FedEx now ships more than 10 million packages daily in over 220 countries.[10]

For me, the most amazing part of his story isn't the outcome or even the incredible company he founded. It's how he got to the idea in the first place. I believe that Fred Smith's idea represents the very definition of creativity: the act of sticking one thing with another in new ways. By connecting how the Federal Reserve processed checks in the late 1960s (a clearinghouse process for an enormous quantity of checks drawn on a large multitude of banks) to the logistics necessary to "automate society," he created an entirely new way of shipping packages overnight that didn't previously exist.

This process of sticky thinking has occurred throughout history. Sam Colt stuck the design of a ship's wheel to the invention of the revolver;[11] Helen Barnett Diserens stuck the concept of the ballpoint pen to a new method of applying deodorant (the Ban Roll-On);[12] and Steve Jobs stuck fashion design to the boring world of personal computing.

Creativity—or sticky thinking—is like a sport in that it requires hard work to perform at a high level. Mastering the necessary skills requires a dedication to practice, practice, and more practice.

Becoming a creative thinker requires the same level of dedication.

Create–Destroy–Repeat: Creativity is a Process

The movie *The World's Fastest Indian* is based on the true story of New Zealander Burt Munro, played by Anthony Hopkins, who took a 1920 Indian Scout motorcycle and modified it through ingenious methods, often using very unconventional and homemade tools. After defying the odds and a number of limitations, he found himself at the Bonneville Salt Flats in Utah in 1967 at the age of 68, breaking the world record for the world's fastest motorcycle under 1000cc. It's a record that still stands today.

In the movie, Burt was asked why he went through all of the trouble to do this at such an old age. His response: "The reward comes from the doing of it." That statement immediately got me thinking about a period in my childhood that has, in many ways, become the benchmark of how I approach life today.

When I was eight years old, my parents bought a new home in a new housing development surrounded by active construction. Being an enterprising young man, I went to each of the construction sites and received "after hours" permission to remove the scrap wood that was piled on the ground, as well as the nails that were dropped on the dirt. I believe most of the carpenters saw this as a way to rid the site of excess "trash," since houses at that time weren't built as efficiently as they are today.

Having watched and learned from the various carpenters constructing the houses, I designed and built all styles of forts including ranches, split-levels, forts with two and three stories, 'A' frames—some in trees and others on the ground. Each time I built one, I would briefly admire it, think about how I could improve it, and then destroy it so I could begin constructing something bigger and better. My mother found this behavior unnatural and frequently said, "You build these beautiful forts, but you never play in them. I don't understand you!"

Prior to high school I was by no means a good student, and I could be somewhat challenging at times for my parents, teachers, neighbors, and nearly every other adult figure. Despite this, most thought of me as "creative," but they never really understood what that meant or how to help me harness it to succeed in school. Without realizing it, they were doing that each and every day by supporting me in my many endeavors, not the least of which was fort building.

In a results-oriented world, the final product or outcome typically gets the majority of the attention and praise. What usually goes unnoticed is the "process" that goes into getting there, which is the primary aspect of creative thinking and ongoing change. Creative people are all about the process, and truly "the reward comes from the doing of it."

Thomas Edison, thought to be an addle-brained youth and most noted for inventing the functional light bulb, had to experiment with thousands of possible filaments before he found one that worked—a daunting task. That success wasn't enough, however. He went on to help design and create a method to distribute the electricity needed to power the bulbs in the first place. He designed the first commercially available fluoroscope (a machine that used X-rays to take radiographs), the motion picture camera, the phonograph, and many more devices that led to him holding 1,093 patents in the U.S. alone. For Edison, it was about the process of creating.[13]

Apple Computer had the best computer available in the late 1970s, the Apple II. However, Steve Jobs wasn't content and pushed Apple forward with the development of the Lisa, and ultimately the Macintosh, before his release as CEO for having been too aggressive on these developments. After his return in 1998, he continued on with his process approach that led to the development of the color iMac, iPod, iPhone, MacBook, iPad, and a variety of innovations in between.

For creative people, it's seldom about the destination. Although the outcomes can frequently lead to wealth and success, the stories of these and a great many other creative people always have one thing in common—the key to their success and all the change that came as a result was the drive and passion related to the *process*, not necessarily the end result.

Bigger is Better

To successfully connect or stick one thing to another, however, you have to be able to draw from a variety of worlds outside your own. In other words, if you're too overly focused within a limited context, you won't have any foundation for comparison. Remember, expand your box! You don't need to necessarily "think outside the box" to make new connections, but you do need to make the box bigger by learning new things, enjoying new experiences, meeting new people, living your bucket list—whatever it takes to expand your horizons while remaining true to who you are. Too often, many of the world's problems can be attributed to people who not only thought they had to "think outside the box," they imagined the box needed to be left behind altogether. As a result, they abandoned their values and the foundation of who they truly are.

Two types of people exist when it comes to creative thinking: people who create new ideas, but do little with them; and people who create new ideas and are able to execute them. Since ideas really only have value if something is done with them, this second type is preferred. Edison once said, "I have far more respect for the person with a single idea who gets there than for the person with a thousand ideas who does nothing."[14]

Edison's statement implies you have to be more than just creative; it is not enough to simply make creative connections. I agree, and I would add two other elements necessary for the realization of a successful outcome: inspiration and innovation. Inspiration is the motivating drive required to initiate creative thinking, and innovation is the process of putting ideas into action, or in other words, using new ideas to add value to existing ones. Edison also said, "The value of an idea lies in the using of it."

Imagineering – The Basis for New Ideas

The term imagineering is derived from combining the words imagination and engineering. Contrary to popular belief, the word's origins were not with Disney. Its roots can be traced back to World War II when Alcoa "invented" the word around 1940 and used it in its advertising to describe their business as "The Place They Do Imagineering." Others such as Union Carbide also used and applied the word to describe their creative activity prior to Disney's adoption of it in the 1960s.[15]

While imagineering has always been used to describe the creation of new ideas, it actually describes a process, not a singular activity—a process that typically begins with **inspiration** (or motivation), moves into **creativity** (connection), and ends with **innovation** (application) (See Figure 2-1).

Inspiration takes on many of the same characteristics as basic internal motivation. Feeling inspired to do something means that future behavior has been given purpose and direction. This purpose usually has to do with meeting an unmet need or solving a current problem. As with any kind of change, people typically aren't inspired to do something until there's a reason.

Inspiration   **Creativity**   **Innovation**
(Motivation)   (Connection)   (Application)

Figure 2-1. The imagineering process.

Almon Strowger, an undertaker in Kansas City in the late 1800s, exemplifies this concept. The wife of his primary competitor served as the telephone operator and worked the cord board at the local telephone exchange. When callers requested an undertaker, or even Strowger specifically, she deliberately directed the calls to her husband, Strowger's competitor. Strowger spent years complaining to his local telephone company, but his complaints failed to solve the problem. As a result, Strowger, who really knew very little about current telephone technology, became inspired to solve the problem himself. The result was the invention of the first automated telephone switch, which allowed callers to direct-dial without having to go through a local operator (See Figure 2-2). His inspiration led to a creative solution and resulted in the redesign of the entire telephone industry.[16]

Many examples exist of people outside a particular "establishment" who become inspired to solve a problem, a problem others have been unable or unwilling to solve or even identify. John Dunlop, a veterinarian, invented the first pneumatic tire in 1887 to provide a softer ride for his son's tricycle.[17] Leopold Godowsky, Jr. and Leopold Mannes, both musicians, invented Kodachrome film in 1916 because they felt cheated after seeing the film *Our Navy*, which was advertised as a color film but had extremely poor color quality.[18] Hedy Lamarr, a popular actress during the 1940s, and George

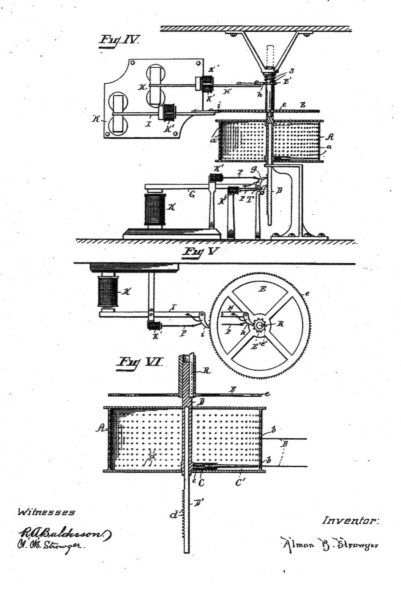

(No Model.)   3 Sheets—Sheet 3.

A. B. STROWGER.
AUTOMATIC TELEPHONE EXCHANGE.

No. 447,918.   Patented Mar. 10, 1891.

Figure 2-2. Strowger's automatic telephone switch design. Courtesy of the U.S. Patent Office.

Anthiel, a composer, developed a secret communications system known as "spread spectrum" technology to help combat the Nazis during World War II. Spread spectrum technology provided the foundation for today's cellular phone technology and other wireless communications.[19]

Indeed, almost anyone can become inspired to solve a problem or meet an unmet need. Although the outcomes aren't always ground-breaking, a person must first be inspired to initiate the creative and subsequent innovative phases of imagineering. We all have been, or will become, inspired at some time or another, and this inspiration may or may not drive us to be creative. However, creativity seldom exists without inspiration first.

Putting Rubber to the Road

Have you ever seen or heard an idea and then said something such as, "I thought of that five years ago" or "I could have done that"? The difference between the idea and the actual realization of it is innovation.

An idea is not enough. You have to do something with it. Innovation is about making new ideas useful—taking the next step to put ideas into action and add value. Edison's idea for the electric light bulb was useless without adequate electricity to power it. For the idea to become useful in everyday life, he had to partner with banker J.P. Morgan to develop a transmission system that allowed electricity to reach homes. The resulting company, General Electric, is now one of the oldest and strongest corporations in the U.S.[20]

The usefulness of an idea isn't always recognized by the cre-ator. In many instances, the ultimate success of an idea may depend on an outsider, one who has been able to observe what the creator could not. The quartz crystal watch was developed in the research labs of the Swiss watch industry, which at the time controlled over 60% of the world market, but Seiko of Japan and the Hamilton Watch Company of the U.S. developed the idea into a market-changing innovation—one that ultimately eroded most of Switzerland's share of the watch market.[21] The engineers at the Radio Corporation of America (RCA) developed the idea for liquid-crystal display (LCD) technology, but it was the Sharp Corporation in Japan that turned the idea into a viable product and something we now take for granted every day.[22]

Innovation also requires a broader, "bigger picture" perspective which allows us to see relationships, systems, and patterns between ideas, thus enhancing their usefulness. The key to effective innovation is often the ability to manage and coordinate many individual ideas into a singular and unified whole or outcome. One of the greatest modern examples of innovation through the management of ideas was NASA's Apollo Program. Sending twelve U.S. astronauts a quarter-million miles to the surface of the Moon required the connected efforts of over 400,000 people from more than 10,000 organizations. To ensure a successful outcome, NASA had to develop the formal structure necessary to orchestrate an immense amount of creativity.[23]

Consider this: a single smartphone today is millions of times more powerful than all of the computers that NASA used during the Apollo Program.[24] Thus, the imagineering necessary to create and develop the vast amounts of technology that didn't yet exist required much more than the simple computers they had at their disposal. They had to use their minds to make connections from among the knowledge, facts, and information learned and stored in their individual mental databases. They were forced to constantly grow the size of their personal boxes. In essence, it was human imagineering at its finest.

Although there's no way to prove this, I truly believe that had these people possessed the smartphone technology we all take for granted, we may not have made it to the Moon. Smartphones have reduced or even eliminated the need to expand our boxes. People today have less need to "know" things since everything is available at our fingertips...we just have to ask for it (literally, just ask Siri). With this kind of access, why take the time to learn anything? Unfortunately, without an adequate mental database from which to draw, imagineering becomes far more difficult.

Where's the Fair?

Tomorrowland is a movie loosely based on the Disney attraction that dates back to 1955. There was a scene in the movie that took place at the 1964 New York World's Fair (a place where Walt Disney featured a number of his new rides and concepts) that got me thinking about the nature of today and what inspires people to expand their minds.

Truly creative and innovative people were once heralded as the rock stars of their eras. People traveled great distances to get a glimpse of Edison's latest invention, the Wright Flyer, one of Tesla's experiments in electro-magnetism, or one of America's first astronauts. They visited World's Fairs (prevalent from 1851-1960s) that were long, two-year events designed to inspire, enlighten, and entertain people from all walks of life. In a single location, a World's Fair showcased and celebrated the world's new ideas and innovations. Compare that with the Iowa State Fair, where the primary focus seems to be the various types of food one can get "on a stick."

Today, people will seldom travel just to see an idea or new invention, and the luster of the World's Fair has diminished along with its frequency and attendance. The last fair held in the U.S. was in New Orleans in 1984, over 30 years ago. Attendance at this fair was less than spectacular—7.3 million compared to the 51.6 million that attended the World's Fair in New York twenty years earlier.[25] While World's Fairs have declined, a growing emphasis is now placed on the tabloid exploits of celebrities, athletes, musicians, and the bizarre worlds of people in social media and reality television.

Neil Armstrong, the first man to step foot on the Moon, passed away in August of 2012. The event could have served as a tribute to both his incredible journey and to all of the people involved who helped make it a reality. However, similar to the decline in World Fair attendance, America's focus has since shifted to other matters, evidenced by the History Channel choosing to air nineteen consecutive hours of *Swamp People* and *Pawn Stars* instead of anything related to Apollo.[26]

Two years later, I had a lengthy conversation with Gene Cernan, Gemini and Apollo astronaut and the last man to walk on the Moon during Apollo 17. During the discussion, he talked about the one thing he was most proud of…and it wasn't walking on the Moon. He was most proud of how he helped to inspire countless young people. Because of his time in the space program, those he inspired went on to accomplish a great many things. However, shortly after expressing his pride in how his efforts had a direct, positive impact on people's lives, he revealed his sadness over how having gone to the Moon no longer inspires young people today. In fact, he sees little else that does, especially in the long-term.[27] Cernan said,

"From an educational and inspirational point of view, the space program is currently failing, which means there's nothing to stimulate the passion of young kids in this country. I have had fifty- and sixty-year-old people say, 'Gene, thank you for what you did. I'm an engineer, a teacher, a scientist

Figure 2-3. Gene Cernan speaking at ciWeek 5 in 2014.

because of Apollo.' We need to rejuvenate that and give the younger generations something to hang their heads on through space exploration." [28]

I believe that for a large and growing number of people the words "creativity" and "innovation" are at risk of becoming nothing more than empty words. They frequently appear in media, schools, television commercials, and presidential campaign speeches, but do they truly inspire someone to action? The Internet is incredible in its ability to make information readily available and accelerate the rate of new advances, but it's also resulted in people becoming lazy in their thinking. Again, why expand your box through memorization and learning when you can quickly and easily look anything up?

Imagineering and the results of the process have always been game changers. They change how people communicate. They change how people travel. They change how homes and businesses operate. They allow people to visit other worlds. They ushered in the atomic age and provided the potential for unlimited energy. They provide food for the growing masses. They add convenience and improve the standard of living for many.

Our society requires more than buzzwords to grow, flourish, and lead. It requires direct calls to action that both motivate and drive people to think differently. We are all responsible for creativity and the subsequent innovations that serve as driving forces for our future, both collectively and as individuals. For those who have come before us, the spirit of their imagineering should be celebrated, and their stories shared again and again to inspire future generations to new creative thought and action.

What inspires you to think creatively and then actually do something with it? How do you help others? In 1955, Walt Disney dedicated Tomorrowland by saying:

> *"A vista into a world of wondrous ideas, signifying man's achievements...a step into the future, with predictions of constructive things to come. Tomorrow offers new frontiers in science, adventure, and ideals: the Atomic Age, the challenge of outer space, and the hope for a peaceful and unified world."* [29]

Like the future envisioned by President Kennedy in 1961 where a quarter-million steps to the Moon are made possible through the imagineering of over 400,000 people, Tomorrowland needs to become Todayland.

Taking a Few More Steps:

• A key to sticky thinking is the natural ability to continually ask questions. Like a little kid to an adult, ask again and again. Question the why or what behind everything. Practice makes permanent. For the next week, ask "why?" about everything in your life. The answers may surprise you. Some inquiries (perhaps most of them) may end after the first answer. However, you may find yourself asking "why?" again and again until some long-term issue or problem gets resolved.

• Another key to sticky thinking is to actively listen to the answers. Start by selecting a few important people in your life and strive to be an active listener with them; your spouse and boss might be good starting points. Remember, practice makes permanent.

• What do you enjoy doing? What about that activity brings you joy? I believe that for most of us, it's the actual act of doing it—the process. Determine what is most important to you and then do more of it. Edison invented because he loved inventing. Jobs pushed the envelope on design and function because that was his passion. Life is short. Spending your limited time engaged in an enjoyable process can be a huge source of happiness in your life.

• Most often, we can be our own greatest Purple People Eater. When do you seem to get in your own way?

• Do you ever think about what you simply accept as truth or fact? Do you do any research to verify your beliefs? The more you dig into the basis of things, the more you will learn, which provides opportunities to make more connections resulting in greater creativity. How are you influencing others? Since creative thinking has been identified by leaders at all levels as one of the most important traits a worker can have, how are you influencing those around you to become better, more creative thinkers?

Fax Me Up, Scottie!

"To boldly go where no man has gone before."

Having watched the Apollo Moon landings on television, the thought of visiting other, distant worlds was the most "groovy" and "far out" thing I could imagine in the late 1960s. Who could have pictured humans traveling at many times the speed of light, "beaming" down to mysterious planets, interacting with other species and humanoids, talking through tiny handheld "communicators," collecting data through small "tricorders," firing "phasers," and actually conversing with a female-sounding computer? Of course, none of these scenarios actually existed at the time, but they seemed very real and quite possible, nonetheless, as I watched reruns of the original *Star Trek* series each week.

Star Trek originated in the mind of Gene Roddenberry, the show's creator. Roddenberry imagined a totally new universe, one remotely plausible, grounded in 1960s science and what was being developed by NASA for Apollo at the time. When Roddenberry conceived a show where humans traveled throughout the universe, he insisted the show be based on solid scientific concepts. Roddenberry and his team would then visualize the 5th, 10th, or 20th generation of what that equipment might look like. He insisted that everything on the show be a logical projection into the future (including the multiracial crew). The show stayed scientifically accurate in those things related to the "known process" of space flight while also remaining accurate in those things necessary to create action, adventure, and entertainment. One notable exception was the rate at which people on earth would age in relation to the crew while they travelled at the speed of light—something with which I'm sure Einstein would take issue. Thought-provoking storylines were relevant to the social issues of the day. *Star Trek* was also the most expensive show of its time and the crown jewel for Lucille Ball and her Desilu Studio. But, with complicated plots and high production costs, the weekly television show suffered from extensive budget and time constraints that forced people to new heights of creativity and imagination.[1]

Roddenberry did more than create a new universe. Like the Apollo Program, he also inspired an entire generation of children to become the engineers, scientists, mathematicians, and researchers who would transform many of the imagined technologies of *Star Trek* into modern-day reality. He inspired women and minorities to take risks and achieve greater success, and he inspired people to envision a world at peace striving to understand its place in the larger scope of the universe.

I was one of those kids. I didn't develop the semiconductor or design the first personal computer, but between *Star Trek* and the Apollo Moon missions, I was inspired to visualize what "could be," to apply my imagination, and to develop my creative thinking and innovation skills. Each of these qualities has enhanced my professional and personal life to this very day.

How is Imagination Different from Creativity?

In the last chapter, I defined creativity as simply a matter of "sticking together" or "connecting" what already exists in new ways to create different or improved ideas. Creativity is part of the imagineering process where an inspired person makes new connections and then makes ideas useful by applying or executing them (innovation).

Imagination is the underlying current that flows within imagineering—like a river that carries a raft forward from point A to B to C (see Figure 3-1). It's the ability to visualize that which may or may not actually exist. It's the ability to *mentally* move a simple concept to a complete idea to a fully realized outcome. Imagination

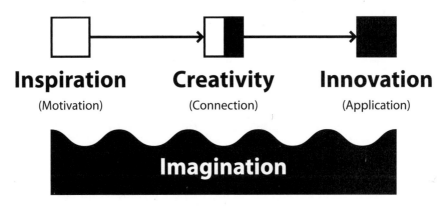

Figure 3-1. Imagination is the underlying current for the imagineering process allowing the mind to visualize.

creates mental visuals of abstract ideas and allows for elaboration of those ideas through future modifications, enhancements, and contingencies. Imagination also requires a willingness and natural desire to visualize differently and take the calculated risks necessary to implement new ideas.

Some argue that it's not truly possible to mentally visualize something without the context of actual experiences or stored mental images. For example, attempt to visualize a red elephant. If you have seen an elephant and recognize the color red, then painting the elephant red (so to speak) in your mind isn't that difficult. For those who have seen the *Star Trek* series or even just its advertising, most could probably visualize a Vulcan or Klingon from memory. However, what if you were asked to picture a Saurian?[2] Could you do it?

Most of us would probably struggle to visualize that particular alien race, unless we had seen (and could remember) the specific *Star Trek* episode that featured it. Without the memories of specific elements, events, or experiences, people often find it difficult to create the mental image of anything outside their personal memories, yet Gene Roddenberry was able to create the image of a Saurian from scratch. How?

Roddenberry understood that the ability to visualize is a necessary tool for effective imagineering. To imagine the first Saurian, he uniquely connected a variety of familiar elements and was able to mentally "see" those connections (the creative process). He was then able to mentally elaborate on those connections, which enabled him to create the storyline and character interaction (the innovative process). Most importantly, he was smart enough to surround himself with other talented, imaginative people who did the same thing. He led his team in a way that allowed imagination to serve as a conduit between creativity and innovation by harnessing the power of visualization. Actually, the orchestrated creativity within the large production team for *Star Trek* was very similar at the time to what was required to put a man on the Moon (program management of three critical, yet slightly different intertwined factors: cost, schedule, and quality).

Every innovation begins with someone asking "What if,"

followed by some imagined concept of the future. This ability to imagine begins during childhood and forms the basis of our ability to visualize as adults. When I built forts as a child, the satisfaction was in the process of creativity, imagination, and accomplishment. I learned construction skills by watching my father around the house and the carpenters at job sites and applied those skills to building ever-better forts. I didn't become an architect or contractor, but this process most definitely enhanced my ability to visualize as an adult. We need to allow our children the opportunity to develop this skill on their own through fantasy play and hands-on activities that foster the spirit of imagination. The same holds true for adults.

Get Out of Normal

I first publicly admitted to being a "geek" during a keynote address on the concept of change. My speech served as an introduction to a presentation by LeVar Burton, who played the character of Geordi La Forge, Chief Engineer on the Starship Enterprise in *Star Trek: The Next Generation*. He also directed about 30 television episodes for the various *Star Trek* series.

A geek and Trekkie like myself has watched all 700 plus episodes of all six television series dating back to the 1960s (multiple times), watched all 13 motion pictures (multiple times), read books like *The Making of Star Trek* and the *Star Fleet Technical Manual*, and even visited the Star Trek Experience at the Las Vegas Hilton back when it was a permanent exhibit.

Like a good Trekkie, I bought a ticket to my first Comic Con in 2015 with only one purpose in mind: to meet William Shatner, the original Captain Kirk. I really wanted a nice picture with him. I wore a sport coat and a nice white button-down shirt—and I *really* stood out. In fact, over the course of the event, I was asked eight different times if I was a security guard. I also realized I was surrounded by nerds.

Now, the difference between a "geek" and a "nerd" is that as a geek, I'm willing to dip my big toe or perhaps even sit poolside with my legs dangling in the water, but a nerd jumps in doing a full-on cannon ball. While nerds have also done all things *Star Trek*, they do it while speaking Klingon and wearing a Star Fleet uniform.

A large percentage of Comic Con participants were deeply involved in cosplay (costume play). I saw one family dressed as crew members from *Star Trek: The Next Generation*—the grandparents were admirals, dad was a captain, mom was a commander and the kids were lieutenants (the fact that I even know these rank insignias adds to my geekness). Bright colors abounded, merchandise changed hands at a furious pace, comic book illustrators had their works on full, brightly-lit display (while they spontaneously created some of the most incredible "doodles" I'd ever seen), active gaming was in play everywhere, and most of all…people were genuinely happy.

While I typically prefer spending hours at the Smithsonian Air and Space Museum or the Kennedy Space Center studying and observing the artifacts from *actual* spaceflight and talking to *real* astronauts, I've come to realize that regardless of personal taste, the key to imagination is the ability to allow yourself to be absorbed into the essence of the moment, to be engaged within the context of your surroundings.

I began the weekend as an outsider who had only ever engaged on the outer fringe of this world. I got a taste of what it was like to completely immerse myself in a unique experience and subculture; one where the primary focus was imagination and the willingness to completely saturate yourself in worlds that don't really exist anywhere except in the minds of the people who created them for comic books, television, and movies. The event was so

Figure 3-2. A group of superheroes at ComicCon protecting
the one odd-looking guy who stood out during the event.

full of energy, motivation, and excitement that by the end, I was sold. Imaginations did, in fact, run wild, and I, too, wanted to be a nerd.

I truly believe that people need to "get out of normal" in order to see things differently, and Comic Con was anything but normal. It was a giant playground of fantasy where you could become anyone you wanted to be, whether it was a superhero, Star Fleet officer, or even a security guard.

We all need a special place to "escape" to open our minds to new things and inspire us to greater levels of imagination and creativity—whether it's heading to Comic Con as a Klingon, jumping on a Harley and heading to Sturgis, engaging in the overindulgence that is Las Vegas, or just allowing yourself to get lost in a good book.

Why Does It Really Matter?

Aside from the inspiration, enhanced creativity, and innovative outcomes, imagination is important for a variety of other reasons.

First, the ability to visualize helps us predict the future. Not in a crystal ball or tarot card kind of way, but in a way that helps people plan and anticipate what might occur at some future point. Through interaction with the world around us, we are able to envision the occurrence of a future event based on similar past experiences and events. The result: our ability (or inability) to visualize possible paths will either increase or decrease the odds of success.

The Power of the Banana

After his college wrestling career ended, my uncle, Terry Paustian, became a high school teacher and coached wrestling for 17 years. Two years after what he thought was the end of his coaching career, the high school athletic director asked if he would consider becoming the coach for the boys' golf team, a decision that was likely driven by desperation; the previous coach had suddenly resigned.

My uncle wasn't a golfer. He knew very little about the sport, and he had a wrestler's mentality—a mindset which is 180 degrees different from those who golf. Reluctantly, he accepted the position.

After selecting the team following tryouts (which he did by simply selecting the 16 players with the best scores), Terry began to watch and observe them during practice. He could tell that many of the players had already established their games through private coaching and the help of others far more qualified. That was a good thing; he didn't feel he had anything of "real value" to teach them that would improve their swing or putt.

Through simple observation, he came to the realization that

golf is a mental game as much as it is a physical one, and his sole job as coach was to teach these kids to keep their minds out of the way of their performance. He decided to apply some of the same psychological techniques he had once used with his wrestlers.

As the team prepared to play for the conference championship—an event the school had not won in over a decade—Terry could see they were all quite nervous. When they stood up and grabbed their clubs to exit the bus, he yelled, "Hey! Who told you to get out? Sit down!" The boys immediately sat. He then reminded them that the school had not won this event in a very long time, but he knew why.

My uncle paused, and said, "In the past, this team has been in it to win right up to the last few holes, and then everyone's game starts to slide. It's not that anyone lacked the talent or desire to win. Your bodies just started giving out after 15 holes. What you need," he paused again for dramatic effect, "...is potassium! It will lift you up, and I guarantee you will play great for the last three holes." He held up a bunch of bananas, began tossing them at each player, and said, "Bananas, full of potassium and natural steroids; you won't believe how this works. Trust me!"

Throughout the match he continued to toss bananas to his players. Before his golfers began teeing off on the 16th hole, Terry looked at them and said, "Banana." They would look at him and nod.

The boys' golf team won the conference championship. They finished undefeated in duals. They won three additional tournaments and the district match, but they only finished second at state. Perhaps it's just a coincidence that my uncle forgot the bananas for that particular meet.

Imagination is frequently the result of a question, in this case the question of how to best coach. Inspired by the need to be an effective coach, my uncle was able to connect an approach used to help his wrestlers with the need to distract his golfers. He was able to help his team overcome a huge psychological barrier to success—themselves. He enabled his team to visualize beyond the moment, beyond the mechanics of their shots, and beyond the spectators. By removing a mental barrier, he helped them relax before each shot was taken and visualize the outcome before it occurred.[3]

You Reap What You Sow

A little imagination can also greatly raise the bar on outcomes and the ultimate rewards generated by them. Gene Roddenberry's original vision of *Star Trek* resulted in the creation of an entire franchise that to date has included six different television series, thirteen motion pictures, a host of fan conventions, and thousands of retail products (not to mention the collectors' resale market). The movies alone have grossed more than $2.5 billion worldwide (adjusted for inflation) at the box office.[4]

The children who became engineers, scientists, mathematicians, and researchers have gone on to turn many of the technologies that Roddenberry imagined in the late 1960s into reality. The cell phones of the late 1990s look eerily similar to the tiny handheld "communicators" used by Captain Kirk and Spock. Today's GPS systems and the voice feature of the Apple and Android phones sound a lot like the female computer voice on the show. The digital writing pads used by Kirk resemble the first tablet PCs. The digital pads that Captain Packard was often seen reading in the *Star Trek: The Next Generation* series (that began in the late '80s) look like today's iPads and Kindles. The video communication between people on the bridge of the Enterprise with those from other worlds or ships seems eerily prescient of FaceTime or Skype. The earpiece that Lieutenant Uhura used at her station on the bridge of the Enterprise closely resembles my Bluetooth headset. One episode from the 1960s introduced color, flat-screen televisions when large, black and white "tube" TVs were the norm. Spock frequently inserted memory cards into bridge consoles decades before memory cards existed. One could even argue that "beaming" has existed for some time now. No, people aren't getting sent down from the U.S.S. Enterprise to some distant planet, but documents are still magically "beamed" through fax machines, and music, art, photos, memos, and a variety of other data are "beamed" daily throughout the world using the Internet. Through the growing use of 3D printers, a variety of tangible items can also now be "beamed" through thin-air using wireless communication.

A Little Pain...Big Gains

The world of *Star Trek* isn't the only place to demonstrate how investments in imagination can yield positive returns and incredible results. Examples can be found in nearly any context:

In 1994, a group of young people used their imaginations to invent the "legend of the Blair Witch," sent three young actors into the woods with video cameras, and created the movie *The Blair Witch Project*. Although the movie's initial production budget was $60,000, the movie went on to gross about $250 million at the box office, not counting the revenue earned through video/DVD sales and sales of merchandise.[5] The movie was listed in the *Guinness Book of World Records* for "Top Budget—Box Office Ratio" (for a mainstream feature film),[6] and since the movie's release, a large number of films have copied its perceived real although actually fake concept.

In the late 1960s, the idea of looking back at Earth from behind Saturn was the stuff of science fiction and beyond the imagination of many. However, in 2006, the Cassini Orbiter (launched in October of 1997) sent back pictures of Earth as seen through the rings of Saturn while the planet eclipsed the sun, thus sheltering Cassini from the sun's blinding glare. A color-exaggerated image was created by the staff at the NASA Jet Propulsion Laboratory by combining a total of 165 images taken on September 15, 2006, by Cassini's wide-angle camera over nearly three hours (see Figure 3-3). The photo was produced by digitally compositing ultraviolet, infrared, and clear filter images and then adjusted to resemble natural color.[7] Similar photos of earth have been taken by other NASA spacecraft, such as Voyager as it left our solar system, but none so incredible or beautiful as the Cassini images.

And, of course, there's the imagination of the two Steves: the late Steve Jobs and Steve Wozniak (a.k.a. Woz). Woz worked at Hewlett Packard (HP) designing calculator chips in 1976. When MOS Technology released its 6502 microprocessor that year for $20, Woz could finally afford the chips necessary to design his own computer. He built the machine and displayed it at the Homebrew

Figure 3-3. View of Earth taken by Cassini through the rings of Saturn.
Photo courtesy of the Cassini Imagining Team/SSI/JPL/ESA/NASA.

Computer Club. Based on the excitement expressed by those at the meeting, his friend Steve Jobs convinced him there was a market for the machine (and also imagined a future for personal computers). Since Woz was under contract with HP, anything he designed while under their employ was technically HP property. After showing the computer to the HP executives and getting laughed out of the office, Woz quit HP and took what would eventually become the Apple I computer to create a new company with Jobs. Through Jobs's incredible foresight and imagination and HP's lack thereof, Apple would become the market leader in personal electronics—especially in the area of design—and one of the most valuable companies in the U.S.[8]

Wow! Just Imagine

During my first computer programming class in college (FORTRAN in 1982), students had to write programs and then store and transport those programs on a series of paper punch cards. Each paper punch card was approximately seven-by-three inches with a pattern of machine-punched holes that represented about 80 characters of data (see Figure 3-4). These cards would be formed into sequentially numbered "decks" to create programs. The card decks would then be fed into a machine that would "read" the cards (the hole patterns) and input the data into a mainframe computer.[9]

As I recently loaded my 32 gigabyte microSD card (about half the size of a small postage stamp) into my digital camera, it occurred to me that, 35 years ago, it would have taken about 400 million of those punch cards (no exaggeration) to equal the data capacity of that tiny microSD card. Obviously, this would have defeated the purpose of having a small digital camera in the first place. To put it in perspective, a stack of 400 million punch cards would be over eight times the height of Mount Everest.

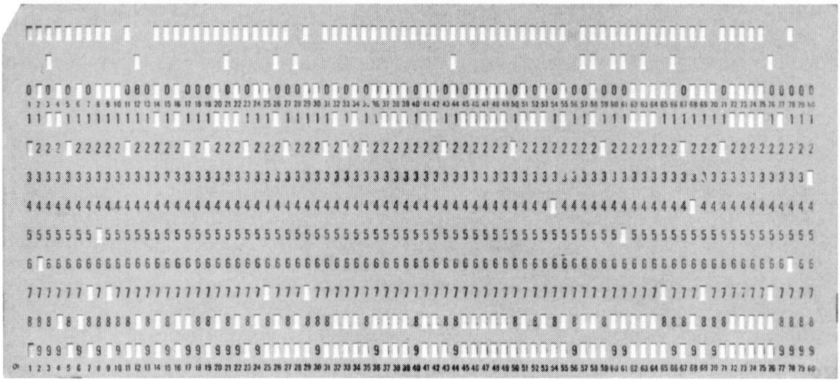

Figure 3-4. A sample of a paper punch card.

My thoughts quickly shifted to Gordon Moore, co-founder of Intel, who was one of the first developers of integrated circuits and silicon chips. In a 1965 issue of *Electronics Magazine*, he authored "Moore's Law," which states that computing power and capacity will double about every 24 months (some believe the number is actually closer to 18 months).[10]

When I contemplate how I once used punch cards to store data and then compare that to what I use today, I can only imagine what wonders await. If we are already seeing many of Roddenberry's ideas from the 1960s coming to fruition now, what lies in the future? If Moore's Law continues to hold true, the computing power we have today will have increased by more than 33,000 times in the next 30 years! Who knows, maybe we will see a future like the one imagined in *Star Trek*—or one even greater—within just a few decades.

Learn to Visualize

I frequently talk with people who profess they aren't very imaginative or that they have difficulty with visualization. I believe the problem is more a matter of "don't" versus "can't." Just as the average person cannot roll off the sofa to run a marathon without proper training, the ability to visualize requires preparation, practice, and the discipline to push your mind off its nice, comfy couch. A number of strategies and activities can improve your ability to visualize, and hence, your level of imagination:

• **Daydream.** In a *Psychology Today* piece discussing studies using brain-scanning technology, researchers have found a significant correlation between robust daydreaming and superior intelligence. Their findings indicate the state of daydreaming was far more active in the "superior intelligence group" than in the "average intelligence group" in those studied. In other words, allowing one's thoughts to bounce around between the past, present, and future while accessing stored knowledge creates stronger memories and experiences. Apparently, those with superior intelligence "allow" this process to occur, thus enabling them to yield greater insights as a result.

Many smart people—from Mozart to Einstein—have credited these insights and their imagination as the source of their intelligence.[11] To enhance the benefits of daydreaming, you must allow time for it, acknowledge it when it's happening, and even take notes for later reference. My greatest daydreaming periods occur while I'm driving, running, and lying in bed before falling asleep. Knowing this, I always keep pen and paper (while others may use their smartphones) nearby in case an especially promising idea comes to mind.

• **Learn New Things.** The process of "making connections" requires an ever-growing base of information and stored knowledge available for the mind to access. Imagination and visualization are no different. Since the ability to "picture" new things in your mind often depends on mentally "altering" things you already know (think red elephants), an ever-growing, diverse base of knowledge and personal experience will greatly enhance your visualization skills. As the marathoner does, hop off the couch and experience the world around you. A mind once stretched by a new thought, experience, or memory never returns to its original shape or dimensions.

• **Learn to Focus.** I'll never forget when I took my daughter to see the IMAX movie *Hubble.* The movie featured one of the space shuttle's missions to repair the Hubble telescope while in orbit and featured a number of the breathtaking images of the cosmos taken by Hubble. Viewed on the immense Omnimax dome screen, the movie seemed larger than life, and it captivated my then 17-year-old daughter's attention. She was engaged and even asked questions—for about 15 minutes—until she received a text message and some new Facebook updates (and then she was mentally gone). In a world today where it seems many of us lack the ability, or the inclination, to focus on any one thing for very long, it's no wonder we struggle to visualize new things.

It's tempting to rationalize this lack of sustained attention by focusing on how much we "think" we accomplish through technology. Sometimes watching my kids, their friends, and the multitude of college students I come in contact with every day is like watching one giant case study in Attention Deficit Disorder fueled by an ever-growing market for specialized energy drinks. Markets typically reflect changes in culture, and a recent scan of these drinks at the local convenience store yielded over 20 brands with names like Monster, Full Throttle, Wired, Nuclear, Amp, Red Bull, and Venom. This behavior isn't all that different from when I was in college, but our options—and the extent to which students relied on them—was limited. Our choices typically included Mountain Dew, No-Doze, and good 'ole coffee to keep us awake and alert. Today, instead of focusing on one thing at a time and doing it well, people often seem to seek whatever boost they can get to maintain their energy levels in order to do many things not quite as well. Learn to focus. (This topic will be discussed in depth in Chapter 7.)

• **Ask.** Years ago, I heard John Falconer, a professor of chemical engineering at the University of Colorado, tell the following story:

> *A student and his professor were backpacking in Alaska when a grizzly bear started chasing them from a distance. They both started to run, but it was clear that the bear would eventually catch up to them. The student stopped, took off his backpack, got out his running shoes, and began to put them on. His professor said, "You can't outrun the bear, even in running shoes!" The student replied as he took off, "I don't need to outrun the bear; I only need to outrun you!"*

I believe the moral to this story is the importance of understanding the true nature of any problem. By questioning the situation and the various options surrounding him, the student was able to adequately define the problem and visualize a viable solution (albeit for himself only). John Dewey, the great educational theorist, once stated that *a problem properly defined is half solved.* When you apply creative thinking—making connections—to a properly defined problem, your odds are greatly enhanced for developing a better, more timely solution.

However, and there's always a "however," properly defining a problem is typically much more difficult than it sounds. It requires asking stimulating, open-ended questions that will facilitate making new connections. It requires asking "Why?" like a child to help us gain greater insight and understanding to the problem. Like the question my uncle asked relating to how best to coach golfers, a question can help us "see" the larger context and bigger picture surrounding any given problem. We are able to get to the core of that problem and better visualize how to solve it by asking questions such as:

- *"Why?"*
- *"What if?"*
- *"What would that look like?"*
- *"Why not?"*
- *"What would it take?"*

A simple question led to the invention of the Polaroid camera, after a 3-year-old girl asked to see a photo of her that had just been taken.[12] A simple question also led a group of watermelon farmers in Zentsuji, Japan, to come up with a more efficient way to ship and store them—the creation of the square watermelon.[13] (The topics of questioning and perception will be discussed in greater detail in the next chapter.)

A Treasure Map of Opportunity

Imagineering provides the mental canvas for creativity and innovation to occur, which allows us to visualize how everything fits together. This underlying ability to visualize how things connect (process of creativity) enables us to elaborate on those connections (process of innovation) in the same mental space. By opening our imagination, the possibilities become endless.

Leadership, personal or otherwise, provides direction and a path towards the successful completion of something that adds value. An active imagination will help guide that process by uncovering opportunities and revealing new worlds previously unimagined or unrealized.

The imagination behind *Star Trek* inspired many people to seize available opportunities:

• Dr. Marc Rayman, chief propulsion engineer at NASA's Jet Propulsion Laboratory, was motivated by the *Star Trek* episode, "Spock's Brain," to develop ion propulsion for deep space probes. Ion propulsion, based on electrically charged atomic particles, is ten times faster than that based on traditional rocket fuel.

• Martin Cooper, a former chief engineer at Motorola and inventor of the cell phone, credits *Star Trek* and its mobile "communicator" for his inspiration.

• Dr. Mae Jemison, inspired by *Star Trek's* Lieutenant Uhura, the African-American communications officer and full member of the Enterprise bridge crew, became the first African-American woman in space. While in orbit in the space shuttle Endeavor, Jemison opened all of her communication sessions with Uhura's famed statement, "All hailing frequencies open."

• Dr. John Adler, a neurosurgeon at the Stanford School of Medicine, admits he was thoroughly inspired by Dr. McCoy's non-invasive techniques used in the Enterprise sick bay. He invented the CyberKnife Radiosurgical System, a computer-controlled laser that ablates tumors and lesions throughout the body. Adler also holds nine United States patents in the fields of surgery, medical imaging, and therapeutic radiation.

• Ed Roberts invented the Altair 8800, a kit computer which

had to be assembled by the purchaser (and led to Bill Gates forming Microsoft). Roberts named the computer after the solar system Altair 6 from the *Star Trek* episode "Amok Time."[14]

And the list goes on.

I even got to see firsthand what could easily be described as a modern-day holodeck as seen on *Star Trek: The Next Generation*. As part of a visit to the Center for the Intrepid at the Brooke Army Medical Center in San Antonio, Texas, I was given a tour of their new CAREN system (Computer Assisted Rehabilitation Environment). The system is a multi-sensory virtual reality lab used for the treatment and rehabilitation of soldiers attempting to acclimate to prosthetics and other mechanical limb replacements. CAREN provides a variety of virtual "action" scenarios (e.g., busy street intersection, path in the forest, etc.) that allow patients the opportunity to relearn how to interact and maneuver within these environments. The system not only included movable walkways and 360 degree dome video, but also sound, wind, and corresponding scents.

What's New Quickly Becomes Old

I'm sure when people read a book or watch a movie or television show that depicts "old society," they struggle with the *how*. How did people survive without plumbing, electricity, automobiles, refrigerators, or simple telephones? Or, more likely, they can't get past how people survived without computers or smartphones. There will be a day, though, that people in the future will look back on us and wonder the same. They actually drove to places in cars? They had to stream video? They talked through smartphones? They only had 300-inch flat panels and used voice-activated remote control?

As it has throughout time, imagination transforms the world at record speed and shows no signs of slowing. If computing power continues to increase at exponential rates as Moore predicted, an active imagination—and the ability to visualize—will remain among the most important and vital traits required to generate desired outcomes.

When Astronomer Clyde Tombaugh discovered Pluto in 1930, it was nothing more than a fuzzy dot at the end of a telescope and remained that way for many decades. In 2001, Dr. Alan Stern successfully led the development of NASA's New Horizon spacecraft designed to learn more about the mysterious planet. After overcoming a litany of barriers including skeptical NASA officials, repeated threats to funding, limitations to the available amount of plutonium necessary to power the spacecraft, and an unforgiving deadline set by the clockwork of the planets themselves, the spacecraft launched five years later and began its three billion mile journey to Pluto—the last planet in the Solar System to be explored and eight months before it was reclassified as a dwarf planet by the International Astronomical Union. Speeding through space—where everything is in constant motion—in excess of 36,000 miles per hour for over nine and a half years, New Horizons arrived at Pluto almost precisely when it was supposed to—a mere 72 seconds early—and began sending back stunning photos of the planet (see Figure 3-5). The success

Figure 3-5. Pluto taken by New Horizons.
Photo courtesy of NASA.

of this mission illustrates how a powerful, positive outcome can be achieved when imagination is mixed with leadership, persistence, and science.[15,16]

Just as the Apollo Program forced us to visualize our planet in a new way by providing us the first pictures of Earth from afar, having gone to the Moon could ultimately seem like a *"tiny* step for mankind"* once we are standing on Mars and other worlds more distant. Imagination is the single most important attribute to creating the future—possibly one thus far unimagined today.

Taking a Few More Steps:

• When was the last time you thought about the "why?" behind the "what," either in your own personal life or in the organization where you work? Taking a look at how you spend your time is a good place to start; our biggest time-eaters tend to be ideas that were once good and have now grown stale. Try to reimagine those ideas and allow yourself the time to mentally elaborate on their ultimate outcomes.

• When do you feel the most inspired to use your imagination and be creative? Where do you feel it? For me, it's when I visit the Smithsonian, Kennedy Space Center, or the Chicago Museum of Science and Industry. Sometimes, it's just watching a great documentary about how something was made or done at the IMAX or on History Channel. Wherever it is, try to do that more often. Sometimes, helping and supporting others' activities may have an even greater impact on developing new ideas. Is there someone you can mentor or support?

• Reconsider your surroundings. How do the people around you and your work/home environment impact your ideas and their outcomes?

• Where and when do you feel most at ease and relaxed? Where and when do you feel most energized and motivated? Wherever those places are, whether real or not, go there often. Whether you need to focus on a project, imagineer new ideas, or solve a problem, the best place to do it is away from "normal."

• Allow yourself time to do nothing except daydream. Go to a relaxing place. Remember to write down all of your ideas and thoughts.

The Great Steffano

Before becoming an Army cavalry scout in the harsh climate of northern Alaska, as well as a husband and father, my son, Steffen, was an aspiring magician. He developed his craft as a young child and by high school mastered many of the skills necessary to amaze and entertain his audiences. I was frequently his audience, as he would test new tricks on me. He believed if a magic trick worked on me, it would work on others.

Frequently, and especially during the early years, I spotted the sleight of hand or figured out the basis for the trick. But, by his late high school years, it became increasingly difficult. One trick mystifies and frustrates me to this day since I have yet to figure it out. My only explanation is something supernatural is going on.

Steffen, or "The Great Steffano" as he often referred to himself, would pull out a deck of cards, fan them out, and show me both sides of the cards to verify their authenticity. He had me pick a card, look at it, and place it somewhere back in the deck, which was shuffled again. He then pulled a clear plastic sandwich baggie from his pocket that contained a single playing card—the joker. I would verify it was the joker and that no other cards were inside the baggie. Next, I'd put out my hand, and he would place the baggie on it with the joker face down. He would then instruct me to place my other hand on top of it.

After 30 seconds or so of dramatic magic stuff (waving the deck over my hands, blowing on them, etc.), he asked me to tell him the initial card I had drawn from the deck. After I confirmed the card, he would ask me to remove my top hand and look at the card inside the baggie, which magically changed from the joker to my card.

To say I've had Steffen repeat this trick for me several times over the years would be an understatement. Each time, regardless of the card I draw, the result is the same. Despite how hard I focus and pay attention to everything happening around me, I come no closer to figuring out the logical basis for it.

Obviously, there is some kind of misdirection: what the eyes see, the ears hear, and the hands touch…the mind delivers. In other words, what I *think* is occurring may not always line up with what is *actually* occurring, which is the basis of perception.

View the images below in Figure 4-1. In the first, a perfect square is placed over a series of concentric circles. In the second, black squares are arranged in a four-by-four grid and spaced the same distance apart. The third is an illustration of a woman.

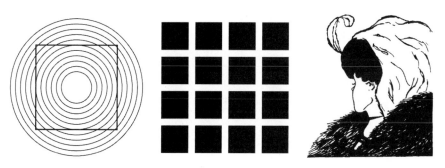

Figure 4-1.

What do you notice? Do the sides of the first square appear to be curved inward? When you look at the second image, do you see shadow-like images where the four corners of each box come together? In this classic illustration of a woman, do you see an older lady or a young lady?

The first two images illustrate how what you see is not always reality. The third image reveals how as humans we can look at the same thing but actually see two entirely different things. Our senses—in this case, our eyes—can play tricks on our minds, which is why it's essential to first try to look at something from as many perspectives as possible prior to passing any judgment. The initial view may be distorted and fail to provide the complete picture.

Have you ever dropped something small on the floor and then had a difficult time finding it? When this happens to me and I start to get frustrated, I remember this varying viewpoint principle and immediately drop to the floor to look across it horizontally—a new perspective that usually yields greater success.

As human beings, it's easy for each of us to view something and come up with different takes on its intent or meaning. We all perceive ourselves and the world around us in ways that reflect our individual values, experience, knowledge, and personalities, based on how we select, organize, and interpret the stimuli.

I once heard a story about a professor who stood before his philosophy class to illustrate a point. When class began, he picked up a large, empty jar and proceeded to fill it with golf balls. He asked the students if the jar was full. They agreed it was. The professor next picked up a small box of pebbles, poured them into the jar, and shook the jar slightly until the pebbles filled the empty space between the golf balls. He again asked the students if the jar was full, to which they agreed. The professor proceeded to pick up a small box of sand

and pour it into the jar, filling the space between the pebbles and golf balls. After asking the students if the jar was now totally full, they agreed with a unanimous "Yes." The professor then pulled out two cans of soda from under the table and poured them into the jar, filling the empty space between the sand, pebbles, and golf balls. The students just laughed.[1]

Creative thinkers know there are many ways to view or perceive something; they become accustomed to stepping back from a problem prior to solving it to see it from as many perspectives as possible. More perspectives allow for more connections and greater opportunities to get creative.

Depending on factors such as age, knowledge, experience, and predisposition (developed as a result of that knowledge and experience), our perceptions shift throughout our lives. Much of this shift can be explained through three specific types of perceptual states: adolescent, experiential, and selective.

Adolescent Perception:
Think like a child

I got to hold my grandson, Emmett, shortly after his birth. As I was looking into his little face while he slept, I thought about how absolutely beautiful he was with his tiny features and more hair than I've seen on my head in 15 years. It occurred to me that life really couldn't be any more straightforward or simple.

As I looked at Emmett, I suddenly realized he literally knew NOTHING, and it was only a matter of time until he started asking the most annoying question that a child could ask—"why?"

I imagined our conversation would go something like this. "Grandpa, *why* is grass green?" "Well Emmett, the green color allows plants like grass to help us breathe." *"Why?"* "The green color is created by something called chlorophyll." *"Why?"* "Well, chloro-phyll is used during photosynthesis." *"Why?"* "Photosynthesis allows plants to use sunlight to turn carbon dioxide and water into sugar, which the plant needs to live." *"Why?"* "Well, when you breathe, you breathe out carbon dioxide which is poison to us, but the grass likes it and uses it to survive." *"Why?"* "So we don't die." Long pause. "Grandpa, *why* is the sky blue?" Sigh. "Ask your mother."

Although it can be frustrating to get the third degree about things we as adults might think are random (and if you're a parent, you know exactly what I'm talking about), this is exactly how children learn, answer questions, and solve problems. This is how they begin to understand the world by making connections and sticking things together in ways that make sense to them. This is why children are so creative. This is also why a child's perception can seem quite literal and as-is. They still don't know all the established and accepted facts and rules related to prior knowledge and experience.

For example, when Steffen was only seven, he and I were at Walmart. As we walked through the store, a new toy caught his eye, and as usual, he immediately had to have it. "Daddy," he said, "can I have this?"

After noticing the price tag, I knew it was beyond the budgetary constraints of a self-employed graduate student. "Steffen, Daddy just doesn't have the money right now."

"But Daddy," he said, "just go to the machine and get some!"

At that time, Steffen perceived an endless supply of cash could be extracted from ATMs everywhere. He had no concept of what it took to earn money and deposit it into a bank. His perception of ATMs was based totally on what he saw—the actual withdrawal. He had no personal experience with earning and depositing, so his perception was the result of his limited experience.

Adolescent perception isn't just related to young children; it applies to anyone, young or old, who lacks the necessary information or knowledge to adequately perceive something in the manner intended. Learning should be, and is, a lifelong process.

In the 1988 movie *Big*, Tom Hanks played a 12-year-old boy named Josh who made a wish he was an adult. When he awoke the next morning, he had an adult body (played by Hanks), but his mind was still that of a 12-year-old. He ultimately found himself working for the development department of a toy manufacturer. Unlike the adults who worked with him, he couldn't help but constantly ask "why?". That question not only caused the company to see great success, it caused the president of the company to show his pleasure with Josh while the other adults at the company took notice (and some became very annoyed).

Unfortunately, asking "why?" is also how children learn the rules in life that ultimately kill the questioning that helped them be so creative in the first place. It's rules like:

> *"Sit still and behave."*
> *"That's not how it's supposed to be done."*
> *"There is no such thing."*
> *"Do it this way."*
> *"Don't color outside of the lines."*
>   (or in my case, *"Don't use a purple crayon."*)

As we age, asking "why?" is discouraged, and over time people stop asking it, conform, and deal with the daily grind of their lives. Ironically, though, whenever we learn about a cool new product or great idea, this is exactly what the people behind them are doing— asking "why?" just like a child. By repeatedly asking "why?", we can get to the core of a problem or situation and true creativity can occur.

I once knew a chiropractor who was not only a professor at the

Palmer College of Chiropractic, he was a master at asking "why?" He told me a story about a patient who came into his office complaining of having constant headaches. When the chiropractor asked "why?", he found the headaches were a symptom of a shifted spinal column which was pinching some nerves.

When he asked "why" again, he found the shift in the spinal column was caused by an unconscious, natural adjustment in how the patient walked in order to compensate for having one leg slightly longer than the other. After being fitted for shoes with a built-in lift on the short side, the patient began to walk normally and the headaches disappeared. Most people today would have just handed him a bottle of ibuprofen, but it wouldn't have solved the problem. Instead, he kept asking "why" until he got to the root cause of the problem.

Perhaps it's time we stop acting like adults and start acting like 12-year-old kids. Maybe it's time to start asking questions more frequently. People might think you're a little annoying, but remember that the intent is to be more creative and strive for better results based on an informed perception—one shaped by knowledge rather than feelings.

In 1961, Kennedy asked, "Why not?", followed by the entire Apollo team asking "How?", "What if?", "When?", and a multitude of other questions necessary to reach the Moon. We had no experience with going there, so the Apollo team had to approach the project like children (albeit very smart children) learning and perceiving for the first time.

Experiential Perception:
Thoughts shaped by experience

The knowledge we gain from learning continuously shapes and

molds our perspectives, which creates a variety of predispositions. Some call this gaining wisdom. However, these predispositions may alter our thinking in ways that narrow our scope or distort how we view things. Others might call this tunnel vision.

Take a moment to solve the following problem, which is typically solved by children in less than a few minutes:

6020 = 3	3305 = 1	8809 = 6	7777 = 0	1970 = 2
2321 = 0	7783 = 2	2022 = 1	3928 = 3	5588 = 4
9999 = 4	1111 = 0	1619 = 2	7175 = 0	7756 = 1
3333 = 0	5395 = 1	6666 = 4	5531 = 0	2253 = ?

Did you struggle with it? Did it create anxiety or stress? I'm sure many viewed it as a math problem, and when you saw that children can easily do it, you may have assumed it would be fairly simple to solve. But why?

Our education taught us that numbers and equal signs reference mathematics, and since adults are so much smarter than children, it must be easy to solve. Yet, the problem has nothing to do with mathematical equations, only shapes and counting. If you count the closed loops in each number, you arrive at the solution. In this case 0=1, 6=1, 8=2, 9=1, and all other numbers equal 0. Therefore, the answer for 2253 is zero.

As-is or literal thinking can lead to incorrect perceptions due to a lack of knowledge (adolescent perception), yet sometimes the culprit of distorted perception can be the facts and rules we learned throughout life. Or worse, it can be due to fragmented perception or incomplete information that results from asking weak questions followed by poor listening.

I was the oldest of three children, and according to my parents,

I was also the most challenging to raise (I like to think I taught my parents how to better raise my younger sister and brother). In fact, my father frequently used the phrase, "You're not listening! That went in one ear and out the other." As human beings, it's easy for us to listen to something and come up with different views as to its intent, meaning, or importance, which in turn leads to different levels of understanding. That's why it's important to ask good questions followed by active listening.

Let's say my wife is in another room, and I ask her, "What time is it?" She may respond by saying, "It's 6:00" or "It's time for dinner" or "Ten minutes later than the last time you asked" or "The same time it is in there." Although all are technically true, they don't get to the same answer in the same manner. Perhaps a better question would have been, "What time does it show on the clock in front of you?" Or, better yet, I could actually get up and check the time for myself.

Effective learning requires not only asking good and appropriate questions, but also actively listening to the answers. Hearing is not listening, and research indicates that most people retain as little as 25 percent of what they hear.[2] Active listening demands intense mental effort to maintain focus while observing and concentrating on the details of what's being said.

Our minds are so busy processing information bombarding us from so many sources like ringing smartphones, text messages, and email notifications. It's easy to mentally move ahead of the speaker, and we may not even realize we are doing so. When you're introduced to someone new, how well do you remember what she said, or even her name? Through active listening, a greater degree of awareness, empathy, and clarity will emerge to enhance perceptual understanding.

While active listening will shape our perception, so too will our experience and the observations made at the time. During my conversation with Gene Cernan, Apollo 17 Commander and the last man to walk on the Moon, he said something that stuck with me.

> *"Today, people's lives are too complicated. They spend so much time dealing with so much stuff they have a hard time appreciating the common aspects of daily life. If going to the Moon and seeing Earth from a quarter-million miles away has taught me anything, it was learning how to appreciate the simple and mundane that we pay little attention to and take so for granted."* [3]

Alan Bean, Apollo 12 Lunar Module Pilot and the fourth man to walk on the Moon, shared a similar view. During a conversation over lunch at a Red Robin, Alan said,

> *"While standing on the Moon—a harsh, unforgiving environment—and looking back at Earth, it caused me to reevaluate my views of everything. I had a newfound appreciation of all things including the most common of them...even the weather. In fact, I NEVER complain about the weather now. I'm just very happy to have it, good or bad."* [4]

My perspective was also altered during that lunch. While sitting with Alan, it occurred to me that no one in Red Robin that day had any idea they were having lunch in the same room with one of only twelve men to have walked on another heavenly body. The experience made me realize how we're all somewhat oblivious to our surroundings. Combine daily clutter with poor listening habits, and it becomes easy to miss the simpler things in life, things that could, and maybe should, provide some of our most important experiences.

Figure 4-2. Earth visible from the surface of the Moon. Photo was taken by Gene Cernan, the last man to walk on the Moon in Apollo 17. Photo courtesy of NASA.

Selective Perception:
Thoughts shaped by incomplete information

In 1999, Christopher Chabris and Daniel Simons performed an experiment to test selective attention. They created a video where two teams of students, one in white shirts and one in black shirts, passed a ball between members of their own team within one large group. Participants who watched the video were asked to count how many times the students in white shirts passed the ball amongst themselves.

Midway through the video, a person in a black gorilla suit walked through the group of students, stood in the middle, pounded its chest, then exited. Study participants were asked, "Did you see the gorilla?" More than half the time, they didn't see the gorilla at all. Study participants were so focused on counting passes between people wearing white shirts that they mentally lost focus on anyone or anything in black.[5]

To test the theory again, Chabris and Simons repeated the study about ten years later. During this test, however, while teams passed the basketball and the gorilla came onto the scene, the curtain in the background changed color, and one of the black-shirted team members walked off the stage. Most people, especially those who had seen the experiment before and were looking for the gorilla, didn't see either the curtain change or the team member exit.[6]

Selective perception can impact our ability to observe something occurring directly in front of us. With so many things happening around us at once, our minds are only able to handle so much information, especially if we are focusing on something specific. Numerous studies test events and corresponding memories—such as studies related to what trial witnesses think and believe they see compared to what other evidence suggests…or what airline pilots see looking out the window versus what their instruments tell them. These studies frequently show that what people think they see doesn't always align with what actually occurs.

Selective perception can affect everyone including the smartest of people. Given an extremely aggressive deadline combined with an excessive amount of required tasks and objectives to reach the Moon, oversights resulted in a fire during an Apollo 1 test killing the crew consisting of Gus Grissom (the second American in space), Ed White, and Roger Chaffee. Despite the fact that over 400,000 people worked on Apollo, it never occurred to anyone that the test—while still sitting on the ground—could be hazardous despite all of the warning signs in plain site.

Despite the tragedy, NASA learned from it. Gene Cernan said,

> *"After the fire, we found a better way of doing it. We pressurized with nitrogen and oxygen and reduced the explosiveness, but at the time, we were concerned with cost and schedules. Although we lost three colleagues and almost the entire space program, the fire allowed us time to create better, more capable, and safer spacecraft, one that would indeed take us to the Moon."* [7]

With the countless demands for our attention today through personal technology, 24-hour news outlets, social media, strategic advertising placements, and more, we tend to pick and choose what we believe is worthy of our attention. As a result, selective perception affects our views and beliefs.

Many years ago, I visited the Museum of Modern Art (MoMA) in New York City, where I came upon a painting that was nothing more than a plain white canvas. Fittingly, it was entitled, "White." I laughed as I thought about the absurdity of this piece. Thoughts like, "They actually hung this here?" and "Anyone can do that!" filled my head as I stood and stared at it...and continued to stare at it. My emotions shifted from laughter to surprise to irritation and back to laughter.

Over the years, I've discussed this piece with many people, often jokingly, but also inquiring as to why and how someone could get away with calling it art. The conversations often shifted to the definition of art and its ultimate purpose. Despite all the great works on display that day at MoMA, I can't remember a single one of them other than this painting. Despite its simplicity, it caused me to think at length about what the artist was trying to do or say. Perhaps that simple white canvas was created so anyone could fill it with imagination, without predetermined limitations. Perhaps it was a metaphor to represent the emptiness that exists in all of us. Perhaps it is nothing more than an ode to simplicity and minimizing the clutter that surrounds us. Or, perhaps it is something entirely different.

I will never know the artist's intent, but I do know how it affected me. When I start feeling overwhelmed with too many things occurring around me, it serves as a personal metaphor to help me mentally regroup and begin again with a clean white canvas, so to speak—a canvas without the clutter, allowing me to select what I will perceive from it.

Ironically, what began as the subject of a personal joke I now see as the most creative piece of all—perception-free, without limitations or constraints, opening the endless imaginations of those privileged to see it, without the selective perceptions or predispositions attached to it.

And, although I know my son wasn't doing anything supernatural with his trick, and regardless of how hard I tried to find the key to the secret, I'm no closer to figuring it out. I realize now I'm only seeing that which The Great Steffano wants me to see, or (and probably more likely) I'm just not seeing the gorilla pounding its chest in the middle of it all.

Taking a Few More Steps:

• Each time you have to generate an idea or solve a problem, try stepping back for a moment. Shift your viewpoint and get a totally different perspective. It may or may not change the resulting solution, but over time you will train your brain to look at every problem from a variety of perspectives.

• The next time you struggle with chaos or feel overwhelmed at a time when you need to be creative, close your eyes and picture a clean, white canvas. Challenge your predispositions by focusing on the simplicity of the canvas and open your imagination to filling it with something new.

• Think back to when you felt the most happy or relaxed. What were you doing? Where were you doing it? What was it about that place that allowed you to experience something positive? Try to identify similar places, both nearby and far away. Use the nearby locations as your go-to "sweet spots" when you want to focus and finish a specific task. Plan trips to your distant locations and set aside that time just to think and process in a relaxed environment. You'll be amazed by the outcome.

• Pick a random object around you. Knowing that there are a number of different ways to view or perceive the object, shift your perspective and describe it in at least three different ways. Repeating this exercise on a regular basis will open your mind to regularly "seeing" problems from a variety of perspectives.

• Don't assume! Ask LOTS of questions—like a young child who constantly asks, "why?" This will often help you get to the root reason for anything, may completely shift your perspective related to it, and help you better understand the cause behind it.

It is Inevitable, Mr. Anderson

In the 1980s, lawn darts were all the rage. Created with good intentions of providing family-friendly outdoor entertainment, lawn darts were in essence foot-long spears tossed between opponents with hopes of hitting the inside of a small plastic ring positioned on the ground.

Scientists estimate the darts hit the ground with over 23,000 pounds per square inch of force.[1] As a result, tossing these steel-tipped projectiles back and forth sent about seven thousand people to the emergency room over a ten-year span, three-fourths of whom were children. In 1987, after a seven-year-old girl was struck and killed by an errant dart launched from the neighbor's house, the Federal Consumer Product Safety Commission (CPSC) reacted and officially outlawed the use of lawn darts.[2] What began innocently enough as a family-fun idea changed into something destructive. Years later, the game changed again, but this time into something much safer—what is now the popular beanbag toss frequently found at football tailgates.

For over 165 years, the Arm & Hammer name has been

synonymous with baking soda and change. During its first century, baking soda was used primarily for what its name implied…baking. As that use declined, Arm & Hammer began marketing its baking soda as a method to keep food tasting fresh by absorbing food odors in the refrigerator. As new refrigerator designs eliminated the need for it, Arm & Hammer changed baking soda's purpose yet again by adding it to a variety of laundry detergents, cleaning supplies, and personal hygiene products. Each time consumer needs changed, Arm & Hammer proactively reinvented itself as a key ingredient in something else.[3]

Anyone who's been alive for any length of time knows that things change. Sometimes change comes as the result of a proactive idea designed to add value to our lives. Other times, change occurs as a reaction to something else. Either way, change happens, whether it's through adding something new and useful or removing something old and unnecessary.

On the other hand, some things never seem to change. The mysterious meat product known as SPAM has sold over eight billion cans of the same basic recipe for over 80 years.[4] Twinkies have been around in their same form for over 85 years, and contrary to urban legend, their ingredients are like any other modern packaged food with a shelf life of only 25 days.[5] By staying true to their core concepts while making consistent, minor alterations along the way, both have managed to maintain a steady path of relevancy despite the changes surrounding them.

So why do some things change while others don't?

I believe the answer to that question can be found in the old axiom, "If it ain't broke, don't fix it." When it breaks, runs its course, or the current outcomes no longer match the desired outcomes, the opportunity for some imagineering finally comes into play. In other words, the driver of change is often the need to create a new solution to a problem when the current, accepted solution no longer works. But, what if something should change, and doesn't?

Some time ago, I was watching an episode of History Channel's *American Pickers*, where the pickers, Mike and Frank, were at a home in California. There was an incredible amount of junk strewn over the property. As I watched them climb through it, I noticed a tiny sign nailed to a tree that said, "Trash is a lack of imagination."

That statement stuck with me. I'm sure it was meant to reference the growing repurposing industry where creative people take one person's junk and turn it into something new and unique. However, it got me thinking about origins and how we have a tendency to give little thought to something once it has been "destined" for its future purpose. This applies to everything, whether it's a tangible object or something as simple as an idea.

For example, I wrote a book about creative thinking published by Simon & Schuster that went out of print in early 2002. Pearson, the parent company of Simon & Schuster, has a policy that states: *"Pearson does not issue royalty checks if the amount due is under $25.00. Earnings under $25.00 will be carried forward to the next royalty statement."* I'm sure the intent was to minimize costs associated with issuing checks for such small amounts, and it probably made perfect sense at the time the policy was put in place.

Since the book no longer generates royalties, I've been receiving the same bimonthly statement for over 15 years detailing how Pearson owes me 52¢. The statement consists of four sheets of multi-colored paper in a 9 x 12 inch envelope, which costs $1.19 in postage (see Figure 5-1). Based on all of the costs involved, including the labor to stuff the envelope and mail it, I estimate they've spent over $600 to date telling me this. Unless someone within Pearson chooses to reimagine the current policy and create a new idea going forward, I estimate they will spend another $2,500 (accounting for inflation) over the next 30 years telling me over and over they owe me 52¢. Odds are I'm not the only author receiving statements like this.

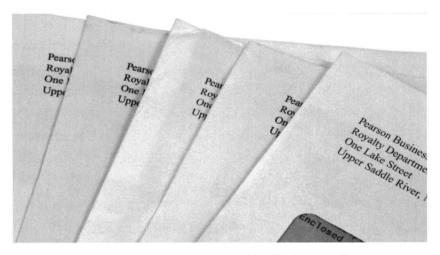

Figure 5-1. Some of the many statements received detailing the 52¢ royalty owed to me.

Red Pill or Blue? Change is a Choice

In the 1999 movie, *The Matrix*, Morpheus, played by Laurence Fishburne, holds out his hands, each containing a single pill: one red and one blue. Mr. Anderson (a.k.a. Neo), played by Keanu Reeves, has to make a choice—take the red pill and free himself from life's current limitations or take the blue pill and return to the status quo.

Life frequently presents us with both red and blue pills. The red pill provides the opportunity to imagine something different and create new ideas that will hopefully result in the change we desire. The blue pill keeps things as they are, regardless of a potential negative outcome. This red pill/blue pill choice is presented to us daily, with varying degrees of importance or outcomes.

When I was an undergraduate college student, a couple of friends and I were driving back to campus in my 1974 Mustang Ghia after a long weekend at home. To say that this particular model of Ford's flagship brand was by far its worst ever would be an understatement. However, I bought it myself, and I was proud to have it (or at least a used version of it).

As we drove down the road, the car began to vibrate ever so slightly. At first, we just shrugged it off as the result of bad pavement. As we continued, the vibration worsened, and it became obvious it wasn't the road causing it. After discussing it, one of us suggested perhaps one or more of the wheels had loosened. So, we pulled over and checked all of the lug nuts—they were all tight. We got back into the car and continued onward having eliminated that idea. The vibration worsened and began to turn into a mild shimmy. We knew something was wrong, but we also needed to get back to campus for upcoming exams.

Like Morpheus in *The Matrix*, two hands were stretched out, each holding a pill that would result in a different outcome. **RED PILL:** Pull over at the next service station and let someone who is qualified examine the car and determine the problem. If the car had to stay for repairs, we could find an alternate way back to campus. If necessary, we could even find an inexpensive hotel room for the night and proactively make some creative decisions as to how we address the current situation. **BLUE PILL:** Press on and deal with it later (the status quo).

We knew in our guts what had to be done, but we chose the blue pill instead. We chose poorly. After driving about 20 minutes or so down the highway, the shimmy became violent, and the stick shift, which I was holding in my hand at the time, disappeared. The transmission had fallen off the car while at high speed, causing an array of serious problems, not the least of which was that we were now stranded in the middle of nowhere.

We are frequently presented with red pill/blue pill moments, and yet we continue to take the blue pill—whether it's continuing to spend money resulting in greater debt, pretending the conflict you have with someone at the office will just go away, continuing to take unnecessary risks when the rewards don't justify them, or wasting time watching your favorite television show when that project sits unfinished on your desk.

Red pill options provide us the opportunity to be creative—to create new and hopefully better solutions to problems that aren't going to go away by themselves. By taking the blue pill with my Mustang, it not only negatively affected my life but the lives of my friends and our parents as they had to spend time and money to rectify what should have never occurred in the first place.

Choosing the red pill can be scary at times because of the unknowns involved, but it's those unknowns that provide us opportunities to become creative thinkers, imagine something different, and make better choices. In *Imagine!* (ironically, the book where I'm still owed 52¢ in royalties), I wrote that imagination consists of two-part thinking: the ability to see an idea in the *abstract* and the ability to *elaborate* on the idea going forward. It's the ability to visualize an idea in the mind before it becomes "real," followed by the ability to visualize the effects and outcomes of the idea after it's implemented.

Imagineering should never be a one-time process. Like with many tangible objects that end up in junkyards because they've "played out" their purpose, a great many intangible ideas solidified into plans, policies, procedures, instructions, guidelines, rules, and a litany of other "ways of doing" continue on into the future, with little imagination or consideration as to how they fit into changing contexts and environments.

For example, in 1986, when data was still downloaded to individual personal computers (think old dial-up connections) and long before people began storing their digital content in the "cloud," Congress created the Electronic Communications Privacy Act. The Act was written in such a manner that characterized any content left on the "cloud" for more than 180 days as "abandoned," thus allowing law enforcement to freely search and seize, without a warrant, any data-based content such as emails. While a warrant is still required to obtain a printed email stored in a file cabinet, such is not the case with cloud-based data that's more than 180 days old. Even though technology has radically changed and advanced in the last 30 years, this law hasn't changed and obviously no longer fulfills its original intent now that cloud-based data storage is common today.[6] Congress will eventually modify the law to deal with current and hopefully imagined future forms of electronic communication, but it will have taken over three decades to do so.

At least on its surface, one would think it would be fairly painless to frequently reimagine the "why?" behind the "what," but it really isn't. People today seldom ever seem content, and they always appear to be trying to find their "happy"—that "something better" in life. The idea of change makes us believe we have power or control over our situations or the circumstances surrounding them. But, I'm convinced that while people in general may actually like the idea of change, they frequently resist it or choose not to accept something different. Combined with the overwhelming amount of daily minutia we all seem to deal with at least in part thanks to technology, it's easy to see how we might revert to our comfort zones and fail to take the time to reimagine anything that's already in place.

Although we may think about changing something for the better, we often tend to wait too long, until we have no choice because the pain attached to the current situation is now greater than the pain of changing it. This shows up in all areas of life—whether it's related to personal finances, relationships, or jobs; or issues with products, services, personnel, or the transmission of your Mustang falling off while driving. By this point though, it's often too late, or at the very least won't result in the most creative or imaginative solution to the problem. Reactive is never a good substitute for proactive.

Always be aware and on the lookout. The danger of failing to periodically reimagine an idea can range from a simple future inconvenience, to spending thousands of dollars to inform someone that you owe them 52¢, to something much worse. Imagineering is a process that should be done daily—not in one day. Waiting until the storm is already upon you isn't the time to batten down the hatches, and it definitely doesn't allow time to develop a good, creative solution to the problem. Change before change is necessary isn't just being proactive, it's often the key ingredient in the recipe for a steady, happy life—like the recipe found in SPAM.

The Power of One

In 2008, singer and songwriter Dave Carroll was flying from Halifax, Nova Scotia, to Omaha, Nebraska, with a layover at Chicago's O'Hare airport. While there, he noticed how the baggage handlers were abusing and throwing guitars around on the tarmac, specifically his $3,500 Taylor Guitar that he wasn't allowed to carry on to the plane. After arriving in Omaha, he discovered it was broken.

For nine months, Dave tried unsuccessfully to have a claim paid on the broken guitar. After exhausting all of the normal and required procedures, Dave resorted to something he knew— music—and created a song and video entitled, *United Breaks Guitars*.[7]

The video went viral and received 150,000 views on YouTube in the first 24 hours, 500,000 views in the first three days, and over 12 million views in about 60 days. It became a public relations nightmare for United. After the first 150,000 views, United offered payment to Dave to make the video go away. It was too late for United, as Dave was now trying to make a point. In the end, Taylor offered two free guitars to Dave, and whether directly connected or not, United's stock value declined by 10% ($180 million) shortly thereafter.[8]

This story illustrates how one inspired person can make a huge difference and create positive change. Dave Carroll's creative inspiration allowed him to singlehandedly take on a huge corporate giant and win. In my various roles in life, I frequently see many people today who truly suffer from a lack of inspiration, the kind of creative inspiration that drove Dave Carroll to create a new song. Therefore, I became inspired myself.

In 2010, we created what would become Celebrate! Innovation Week (or ciWeek) at the West Des Moines campus of Des Moines Area Community College. Short of personally taking on a corporate giant, I feel the best approach to inspire others is meaningful storytelling through direct interaction with the people who *are* the stories—current, living creators of new ideas and the latest innovations. Through direct engagement with the "who behind the what," the stories come alive and can have a direct, emotional impact on those fortunate enough to hear them.

Through our annual ciWeek, one week each year is set aside to provide students and the community as a whole the opportunity to directly engage with people (some famous, all inspired), who have dreamed, created, and accomplished. It's a thought-provoking and highly interactive week that lets attendees listen, absorb, and engage with people who, under normal circumstances, they wouldn't have the privilege to meet. The event is entirely paid for by a number of generous sponsors, making it free to all who attend.

Previous ciWeek presenters have included four of the men who have been to the Moon; the father of the personal computer; television personalities who focus on science, invention and ideas; explorers who have been to the wreck of the Titanic and the furthest depths of the ocean, to the highest mountain peaks and most dense jungles; engineers who are developing the growing commercial space industry; inventors of incredible bionics, robotics, and animatronics; Academy Award-winning visual effects creators and animators; #1 best-selling authors of both fiction and non-fiction; nationally known artists and even connoisseurs and creators of wines and cheeses.

People frequently ask me why we invest the large sums of both time and money to make this happen every year. It's because following each event, a wide variety of people personally share how the experience has had a direct, positive influence on them and has changed their lives.

It only required the vision of John F. Kennedy to start taking a quarter-million steps to the Moon, Steve Jobs to begin Apple

Figure 5-2. Steve Wozniak, co-founder of Apple, speaking at ciWeek 3 in 2012.

Computer, Henry Ford to develop a new method of production to bring automobiles to the masses, Jonas Salk to create a vaccination for polio, actress Hedy Lamarr to invent spread spectrum technology (which became the basis for today's cell phone technology), Fred Smith to envision a world with overnight shipping, and Gene Roddenberry to imagine a technological future in *Star Trek* that inspired others to bring much of it into today's reality.

Thousands of people are touched each year by ciWeek. Any one of them could be inspired to create or invent something new to change our lives for the better. *Isn't one enough to make a difference?*

Yellow is the New Blue: *Socially Influenced Change*

Change, in most of its forms, is typically the result of either a proactive or reactive action. In other words, when someone or something causes or creates change, it was the result of either an intentional effort or because something else occurred first, thereby requiring it.

I once taught a graduate class in marketing comprised of working adults who were somewhat older and wiser than typical college-aged students. There was one exception—a 22-year-old who showed up to class at least 15 minutes late, week after week.

Jim (not his real name) always quietly came into class, sat at his regular seat, and attempted to determine what he missed, occasionally disrupting one of the students sitting next to him by asking questions.

We reached a point in the semester when it was time to discuss the psychological effects of product attributes and the power of influence. To illustrate the concepts, I brought in a large yellow boom box (1990s slang for a portable radio with large, visible speakers), sat it on the table in front, and explained to the class how the color blue affects perception, mood, and purchase choice. I explained that I wanted to use an entertainment-related product, but the only one I had available to show as a visible prop was in yellow. I asked the class to "pretend" and "imagine" the radio was actually blue. I told them the color was critical to our discussion and implored them to always refer to the radio as blue. They all agreed.

Our discussion began and everyone played along. Eventually, Jim arrived, late as usual. He sat down and began listening to the discussion. "Will the radio's blue color have a positive net effect on sales?" "What shade of blue are people most drawn to for outside activities?" "What if the radio was a darker shade of blue?" I watched the expression on Jim's face change as it quickly became apparent that what people were saying didn't jive with the visual object before him.

Jim whispered to the student on his left. The student looked at him and said, "Shhhhh." Jim looked puzzled, squinting his eyes at times. For a while he even looked angry, but that ultimately turned to concern. After an hour and a half of discussion, the class concluded.

The professor whose class was scheduled for the room next was standing in the doorway talking with a student. She knew what I was doing and was prepared for it. Out of the corner of my eye, I watched Jim get out of his seat, walk over to the professor, and ask her about the color of the radio, to which she replied, "Blue." Jim then walked over to me and asked, "Professor Paustian, if somebody is colorblind, how do they know?" I responded, "Why do you ask, Jim?" Jim just shook his head.

I smiled and explained to Jim how he was set-up as a result of being late to class. After we both laughed a bit (mostly me), I told him I was going to use this as a teaching point related to the power of influence at the beginning of the next class. Jim was never late to class again.

Whether Jim truly believed he was colorblind or not, at the very least, he began doubting himself and what his eyes were telling him. We are constantly influenced by others, whether it's advertisers trying to mold our views on a particular brand or product, a politician trying to gain support, or a friend or co-worker trying to sway you to his or her point of view. Unlike the conscious "red pill/blue pill" decisions we make, these influencers frequently guide our decision-making and cause changes without us even realizing it.

Sometimes it's easier to accept something as fact rather than taking the time to verify its authenticity. While I acknowledge there are some things we have to accept on faith as there is no way to tangibly verify its truth—such as a belief in God—the act of researching, studying, and learning is an opportunity to build our mental databases. This ultimately provides more material from which we can make connections and become more creative.

Influence is also a two-way street. The more you know and learn, the more you will be able to influence and lead others toward desired change, better outcomes, and higher levels of success.

Sweet Dreams: *Change and the Subconscious*

I once took a long road trip with my college girlfriend. There was little that I didn't like about her except for one irritating behavior. Whenever we took a long drive, she almost instantly fell asleep while I sat in silence listening to the radio.

It was spring break, and I happened to be in the middle of a very interesting psychology class where we were learning about the subconscious mind and how it affects our behavior. So, like any young, naive college kid trying to apply what he had learned, I decided to conduct an experiment.

I picked a random item—in this case the color green—turned

down the radio, and softly whispered into her ear, *"You HATE green. Green is EVIL. Green is BAD. Green causes PAIN."* I then turned the radio back up, waited a couple of minutes, and repeated the entire process.

After about thirty minutes, I changed the dialogue. *"You DON'T LIKE green, green causes you immense PAIN, green is the favorite color of the DEVIL, green is UGLY."* Like before, this went on for about thirty minutes.

I then shifted the dialogue to something positive. To begin, I purposely picked a color I knew she would never pick on her own and probably didn't even really know what it was—indigo. *"You LOVE the color indigo. Indigo is HAPPY. Indigo is SWEET. Indigo is PRETTY."* This went on until she awoke.

After giving her time to fully awaken, I decided to spark up a conversation. "Sweetie, when you were asleep, we drove by a ton of green grass." She didn't respond, but I could see a slight frown. I then followed by saying, "If you could pick any color for grass other than green, what would it be?" Without any hesitation and a smile, she responded, "Indigo."

In contrast to the *conscious* social influence that affected Jim's thinking while in class, what about something as simple as what enters your *subconscious*? When I asked why she chose that color, she had no idea why. My experiment, however twisted, was a success. I confirmed what I had learned in class—our subconscious never sleeps and picks up information 24/7.

Have you ever woken up to find someone standing over you who hadn't made a sound walking into the room? If you consciously set your alarm to wake up the same time everyday, do you find yourself waking up on your own before the alarm sounds? Our subconscious serves as a protection mechanism. It's aware of our internal clock. And it takes in information that affects our conscious thinking.

Have you ever been driving down the road listening to the radio and a 20-year-old song comes on that sends you down memory lane? I believe our minds are like immense storage devices that retain everything—every event, every smell, every sound, every feeling, everything. However, retention is not recall, and recalling past information can be a bit of a challenge.

Back when all of those events were taking place, odds are that particular song was frequently played on the radio. Your subconscious connected the two, and the song became connected to the address in your brain where those memories are located. Those memories came flooding back because of the subconscious connection to that song.

While the subconscious can have an unintentional influence on our thinking, it can also be a powerful tool to recall information and solve problems. After spending time trying to solve a problem without success, have you ever moved on, only to have a random "Aha" experience about it later? That's your 24/7 subconscious working the problem behind the scenes.

Proactively using our subconscious mind not only helps with memory, it can also help create solutions to problems that our conscious thinking can't address. Even Albert Einstein once said, "Why is it I always get my best ideas while shaving?"

New Isn't Always Better

I once had the opportunity to observe professional drag racing firsthand at the NHRA Nationals in Brainerd, Minnesota. As part of the experience, I got to walk the U.S. Army Top Fuel car to the starting line and stand behind it as it launched down the track. What I didn't expect was to be physically knocked backwards by the shock wave created by the 8Gs of force generated when the car took off. I couldn't see the shock wave, but I definitely felt its power.

Change, too, can be difficult to see, but its effects can have a profound impact. The ability to see either something that doesn't yet exist or the oncoming effects of change requires imagination, or the ability to mentally visualize and elaborate on abstractions.

Someone once imagined a future where food is prepared in *Star Trek*-like replicators, humanoid robots walk and interact with people, and 13-year-old gamers work to cure cancer. As a result, 3D printers are now able to print edible food,[9] and the Robotic Challenge through DARPA (Defense Advanced Research Projects Agency) has created androids that walk and move like humans.[10] Speaking of DARPA, they run a public computer game through social media called Foldit where young gamers try to fold proteins, one of the most difficult biochemistry barriers to curing disease.[11]

Imagination is a fundamental trait of an effective leader, both of others and ourselves. While history is loaded with people who lacked this invaluable trait, successful organizations are most often led by those who have vivid imaginations. These people are able to see where the world is heading and develop products and services that anticipate the transformative change (think Steve Jobs, Bill Gates, and Elon Musk).

While it's often tempting to react to change and make rash moves to adapt to it, newer doesn't always mean better. Change isn't always transformative. For example, following the release of Windows 2000, it quickly became apparent the upgrade wasn't an improvement and most people reverted back to Windows 98. As a child, I never had to wait for *Gilligan's Island* to buffer, and the battery in my HP-12C calculator purchased in 1988 lasted for 25 years of regular use before it had to be changed (seriously). Today, I'm lucky to get a few hours of battery life out of anything.

When new, "game-changing" ideas are introduced, people are

frequently quick to jump on the proverbial bandwagon, often with what seems like an adapt-or-die mentality. They incorporate the ideas into their own existence with the hope of remaining relevant—easy to see when looking at the rapidly changing landscape of products such as smartphones, tablets, app development, and content distribution over the last decade.

While many of these new ideas are quickly incorporated into current ideas and products with the intent of making them more efficient, capable, and feature-laden, sometimes the opposite can occur. For example, appliances were once designed for simplicity of function. Refrigerators could easily last for decades, and older units often ended up in garages and basements for beverages and other items. Recently, I had to replace our three-year-old, $2,000 refrigerator because of a coolant leak. The unit was designed with multiple sealed cooling systems and state-of-the-art, high-tech electronics. Because of the complicated design, two different repairmen told us it could easily cost about $1,500 to repair it. We just bought a new one.

However, when the world does embrace change, it can also leave opportunities behind. Despite trying to fully utilize a Palm PDA in the early 2000s and then a smartphone, I'm still more efficient and successful using an old-fashioned paper planner—and new, stylish paper planners continue to line the shelves of many retailers every year. Speaking of paper, the tablet computer and eInk readers were supposed to mark the end of the traditional book, yet paper book sales are as high as ever and eBook sales declined almost 13% in 2015.[12] Despite desktop publishing and an array of high-quality, low-price color printers, print shops using archaic letterpress machines—those that use wood and steel type—are popping up all over. After years of improving food production through genetic engineering, food trends are giving way to organic and old-fashioned, farm-to-table production.

While I continue to be a huge advocate for the creation of new ideas and awareness of the possibilities those ideas bring, I believe it's more important to stay true to your own values, core competencies, and the passion that fuels them. New and different isn't always better. Sometimes, as those new ideas shape the world around us, it requires imagination to see how our old-school ideas can become new-school thinking.

Work HARD, Not SMART

There's a frequently-used acronym related to creating goals— SMART—which stands for Specific, Measurable, Achievable, Realistic (or Relevant), and Time-Based. As a college professor, I taught for over 20 years a variety of concepts required to attain a desired goal or future vision. I frequently discussed the importance of creating SMART goals and how they were absolutely critical in order to accomplish this desired end.

And it was hogwash.

It's not that SMART goals are necessarily bad, but I now believe they're flawed if what you are trying to achieve requires a behavioral transformation or major proactive change. Specific, Measurable, and Time-Based are all fine attributes and should automatically be built into all goals. It's the Achievable and Realistic (or Relevant) parts I've been struggling with for some time now, especially after reading a piece in *Forbes* discussing how SMART goals can be dumb.[13]

In the writer's opinion, both Achievable and Realistic actually act as impediments to creativity and don't really enable genuine movement or progress. I completely agree. Those attributes smack of phrases like, "Don't bite off more than you can chew," "Stay within your available resources," "Be careful what you wish for," "Play it safe," "Don't do anything stupid," and "Keep your eye on the ball."

Because of the Realistic (or Relevant) attribute, Yahoo decided to pull out of purchasing Facebook during its early years because of an overreaction to a short-term market dip[14] (an article in *CNN Money* now predicts Facebook will at some point have a $1 trillion valuation[15] while Yahoo continues to suffer); the company Digital Research passed on partnering with IBM for the creation of an operating system (after Bill Gates sent them there in the first place—the result was Microsoft creating MS-DOS and Digital Research is now long gone);[16] Kodak waited way too long to embrace digital photography because it didn't require actual film (and subsequently filed bankruptcy); and the list goes on.

A great many poor decisions have been made by people based on "unrealistic" or "non-relevant" views—views often rooted in how things currently are as compared to where they will or should be because they were unable to imagine something different.

Thankfully, John F. Kennedy didn't listen to the pundits in 1961 who claimed going to the Moon was unachievable. If NASA had used SMART goal thinking in the 1960s, we would have never gone to the Moon, especially within the "unrealistic" context at the time: less than 9 years to complete, over $25 billion cost (about $144 billion in today's dollars), and less than 20% of the necessary technology required to do it. Also, over 50% of the country didn't even want to fund the project, especially since we were involved in the costly Vietnam War. But, this project was absolutely necessary and relevant in order to stay ahead of the Soviet Union.[17] As part of a 1962 speech given at Rice University, John F. Kennedy proclaimed:

> *"We choose to go to the Moon. We choose to go to the Moon in this decade and do the other things, not because they are*

Figure 5-3. President John F. Kennedy speaking at Rice University. Photo courtesy of NASA.

*easy, but because they are **HARD**, because that goal will serve to organize and measure the best of our energies and skills, because that challenge is one that we are willing to accept, one we are unwilling to postpone, and one which we intend to win, and the others, too."* [18]

The result was not only one of the greatest achievements of mankind, but also the staggering development of thousands of products and industries (such as microwaves, water purification technology, polarized and scratch-resistant lenses, lithium batteries, kidney dialysis, NASCAR cool suits, solar panels, and companies like Intel[19]), not to mention the inspiration of an entire generation and the creation of worldwide optimism during a difficult time in our history.

My point is this: any goal that requires *transformative thinking*—thinking required to change deeply entrenched behaviors, habits, and modes of thought—isn't SMART. It's HARD.

HARD goals require a total change in thinking—a recognition that transformation is difficult and realized through intense focus, effort, and tenacity. Our greatest accomplishments in life aren't easy. Therefore, I believe HARD goals include the following attributes:

• **HONEST:** One of the biggest reasons that many goals are never achieved is because people do not honestly, deep down, desire them. Often, they are the goals of someone else, such as a spouse, parent, physician, supervisor, or the organization as a whole. They could also be the goals that have been deemed "good" by a majority of society. Regardless of origin, unless they are your goals—goals you totally believe in, desire above all else, and have built into the very core of your being—they will never be realized.

• **ACTIONABLE:** These goals must be something you can begin taking immediate action toward…not someday or at some future point. If these goals require other things to occur before you can begin working on them, the likelihood of success is diminished. The Apollo Moon program began the moment Kennedy shared the goal with Congress.

• **RADICAL:** Most goals related to change frequently require some form of radical or significant shift in behavior. Typically, a minor or slight change in thinking is inadequate to achieve transformative change. For example, effective weight loss requires a sustainable, *permanent* change in diet from what you previously considered normal, acceptable eating. Eliminating debt requires a sustainable but *substantial* *shift* in your spending patterns, purchasing behavior, saving, investing, and all of the other ways you think about money.

• <u>D</u>ETAILED: HARD goals require a plan of attack. This is where desired action is specifically detailed, time-based, and measurable. This plan of attack should be incremental in nature since the power of progress is typically found and achieved through daily activity. Again, NASA had an extremely detailed plan with a series of very specific, short-term objectives required to land on the Moon in less than nine years.

Research shows that fewer than two out of ten employees strongly agree their goals will help them achieve great things, and even fewer strongly agree their goals will help them maximize their full potential.[20] Goals that truly lead to transformative change, the kind of change that will help you achieve great things and maximize your potential, require a HARD focus. A desired goal must be *honest* and true to both who you are and where you want to be; it should be immediately *actionable* and not something you have to wait on if you're ready to go now; it should be *radical* in that to get there requires a sharp shift from the behavior that's obviously not working now; and it should be *detailed* so it's extremely clear as to the steps, resources, and time required to achieve it.

A HARD goal doesn't necessarily mean it's extraordinarily difficult to accomplish, but it does require a higher level of intensity to achieve the desired transformation than a SMART goal. If you truly want to proactively change something in your life, you will need to change your mind—a change in thinking that directly affects your daily behavior.

Five Minutes of Absolute Terror

As someone with an intense fear of open-air heights, I'm not exactly sure how I rationalized standing, tethered to a "pilot," in front of the open door of an aircraft a little shy of three miles off the ground. Although I took slight comfort knowing the pilot was one of the best in the world—a U.S. Army Golden Knight with over 9,000 jumps—I seriously questioned my decision in that moment.

Perhaps I wanted to prove to myself I could let go of my fear or, maybe, I felt baited into it by my Army friends who kept calling me a wimp, among other things. Whatever the reason, there I stood, terrified and mentally frozen with my heart pumping like it would explode in my chest.

I had no choice but to trust my pilot, a man half my size. As I hung out the door while he held on to make final preparations, I resisted looking down for those few seconds as I had absolutely no control. I tried to ignore the sudden urge to clutch something—anything—to save my life, especially since grabbing something at this point could cause serious injury.

The pilot tapped my shoulder indicating we were about to jump. After three forward lunges, we began our free-fall descent from over 14,000 feet at about 120 miles per hour. Breathing was difficult, and my cheeks flapped from the massive intake of air. Because I was traveling so fast at such heights, I didn't realize I had begun flapping my arms like a large, prehistoric bird. Perhaps it was the lack of context. When I saw the curvature of the earth, the ground looked like a blurred mass of color and undefined features. At 5,000 feet, the pilot deployed the main chute. Our speed and descent slowed, which allowed us the freedom to circle, twist, and glide as the pilot wished.

Then came the most terrifying question I'd ever heard: *"Would you like to take the controls and fly the chute?"*

When the main chute deployed, I had clenched my harness for dear life, and the decreased speed and increased clarity of detail on the ground below reminded me of my fear of open-air heights. No way was I going to let go and grab the steering controls. Letting go of the harness would have meant abandoning my false feeling of security. I was convinced if I let go, I would certainly fall to my death.

Of course, that thought was ludicrous; I was skydiving with one of the best in the world. Yet we all struggle with "letting go" of what feels safe at times, whether we're clinging to unnecessary fears, flawed thinking, insecurities, bad habits, or even a parachute harness.

At its very essence, creativity is typically at the center of change, which often brings about a variety of emotions in people, not the least of which is fear and all the "What if?" scenarios that

Figure 5-4. Holding the harness that made me "feel" safe while we plummeted toward the ground at a speed of 120 MPH.

come with it. Fear is typically a function of the unknown, and our inability to let go of it keeps us from taking proactive steps toward change, experiencing new things, or taking advantage of new opportunities—in this case the opportunity to control the chute and the direction we were taking. Once we are able to take that first step forward, however small, the unknown becomes a little less so, and each step thereafter builds confidence to take the next.

Change, in all of its forms, is inevitable. Whether it's from a proactive choice between two pills like in *The Matrix*, the influence of others, a product of our subconscious, or an outcome of working HARD, change will happen. What determines the impact it has on our lives is our ability to visualize it before it becomes real.

While the skydiving experience didn't cure my fear of open-air heights, I did grow as a result. If nothing else, I took another step (albeit almost a three-mile one) toward facing—and letting go of—my fears, the first step toward visualizing positive change and one of many steps to come.

Taking a Few More Steps:

• Do you have a tendency to keep making the same mistakes over and over? Are you doing things that you know in your gut you need to change? Stop and take a pause. Take out a piece of paper, and at the top write a statement of the issue or problem at hand. Below that, create two columns with the words "Red Pill" and "Blue Pill" at the top of each column. Start with the Blue Pill column and create a list of probable outcomes if you continue treating the issue or problem the same way. Once complete, create a list in the Red Pill column of specific changes that could be made to address the issue or problem along with any and all possible outcomes resulting from each. Once done, take an honest look at both lists. Make a better choice than the one I made with my Mustang. It may not be easy, but it will ultimately yield a better outcome.

• Your dreams are frequently the result of your subconscious trying to process both old and new information, which is why the buzz of the alarm clock can be interpreted as a fire alarm in your dream. However, dreams also provide a place for great solutions to problems to come to the surface. To harness those ideas, try this: each morning when you awaken, immediately write down anything you can remember from your dreams. Over time—and that time will vary among people—you will train your mind to be able to freely recall the details of all the dreams and ideas you had the night before. It could be a game changer for you.

• Occasionally when the world around you shifts, it creates an opportunity. Can you reapply some "old-school" ideas or practices that made your company or organization great in a new-school way? Perhaps it was people-centered, very personal customer service as compared to using overseas or automated service, or perhaps it was the laborious, handcrafted approach to production used to make your products as compared to new state-of-the-art production techniques. Whatever it was that differentiated you and formed the basis of who you are today could become the new-school idea that launches you into the future.

• What do you fear? What keeps you up at night? The next time you find yourself up against it, take a baby step. Challenge your fears one at a time by continuously reducing the unknowns surrounding it. The more you know and the more experience you have with it, the less you will fear it going forward.

• What do you really want to do or accomplish? Is it something you can begin right now? What kind of radical or major shift will be required as compared to what you're doing now? What will it specifically take to accomplish it and how will you measure progress? This type of HARD goal thinking can dramatically impact your life if taken seriously.

What's the Big Idea?

Ideas come in all sizes, shapes, and flavors. Ideas can range from great to good to bad to downright horrible, and through the Internet, social media, and the many other 24/7 sources of information in existence today, ideas have also never been so plentiful. They're everywhere, and there's no shortage of them. But, how do you know if an idea is any good? What constitutes whether something is actually good or bad? And what are the characteristics of "goodness?"

Was Apollo a Good Idea?

Most people today would consider the idea of having gone to the Moon a great idea, but that idea is being viewed in hindsight with the luxury of knowing its ultimate outcome. When Kennedy first set the goal for the country, there were a great many people who thought the idea was absolutely terrible, especially in light of everything else the country was dealing with at the time. Also, one has to realize the idea of going to the Moon was the culmination of many smaller ideas, not all of which were very good.

For example, testing Apollo 1 at sea level in a high-pressure, 100% pure oxygen environment became a bad idea when a frayed wire shorted causing a fire in the command module which killed all three astronauts—an idea that almost resulted in cancelling the Apollo Program altogether.[1] The idea to have two separate contractors design and build the CSM (command service module) and the LM (lunar module) resulted in two distinctly different carbon-dioxide scrubbers for each spacecraft (one round and one square). The result required some nifty, quick fix imagineering to convert one into the other using materials the astronauts had at their disposal during Apollo 13.[2] Without this fix, the astronauts would have asphyxiated.

Probably the worst idea of all was killing the Apollo Program in 1972 in favor of the very costly Shuttle Program. Apollo could have continued on for many years, evolving to achieve new goals at a relatively low cost since the technology was already designed, tested, and manufactured in facilities created to do so. Instead, NASA decided to toss the Apollo investment and start over with a wholly new, very complicated Shuttle Program, a decision that was incredibly wasteful both in terms of learned capabilities and money.[3]

Ironically, NASA now relies on the Russian Soyuz capsule (1960s technology) to shuttle our astronauts to the International Space Station and also relies heavily on the private space sector to get supplies there, primarily using the SpaceX Falcon rocket with a capsule similar to that of Apollo.[4] NASA's new Mars Program design includes a souped-up Apollo capsule called Orion—something even NASA admits on its website.[5] Imagine how far along we'd be in both technological advancement and money savings had they continued development of the original Apollo technology.

So What is a Good Idea?

When considering whether an idea is any good, a litany of questions comes to mind:

- Does the idea solve a problem?
- Does the idea meet a genuine need or want?
- Does the idea add value to another idea?
- Does the idea do something different and better?
- Is the idea even possible?
- Does the idea serve its defined purpose?
- Is the idea tested, proven, or validated?
- Is the idea supported by a majority of others?
- Does the idea have a clear purpose that resonates with people in an emotional way?
- Is the creator of the idea personally vested in it?
- Is the creator of the idea passionate about it? Do they truly believe in it?
- What are the short- and long-term consequences of the idea?
- Are people better off with the idea?
- Do people you trust or don't know support the idea?
- Will people pay for it?
- Does the idea have an acceptable cost versus benefit?
- Does the idea make life more convenient, efficient, or affordable?
- Can you explain the idea in the simplest of terms?

This list could easily go on and on. Perhaps the best way to determine whether or not an idea is good is to first look at what "good" means.

The Process of Good

Good is difficult to define, but one thing is fairly certain: most people tend to have feelings about what good is and seem to know it when they see it. Therefore, I believe that to define "good" is to actually define a process, a process of evaluating the relative value of something as compared to something else in a similar context or category.

I believe this "value-sorting" process is the very essence of determining or evaluating the goodness of an idea. It becomes a personal process whereby a comparison is made in the mind of the evaluator between the idea and all other competing ideas based on some established criteria. These criteria can be divided into two basic categories: *musts* and *wants*. The musts set the initial bar for comparison and can include a variety of "required" characteristics—characteristics such as cost, marketability, timing, affordability, convenience, simplicity, etc. All potentially good ideas have to meet the musts to even be considered. Wants, although not required, are "desirable" characteristics and are used to differentiate amongst the viable ideas that have met all of the musts.

For example, you have decided to purchase a vehicle, and your primary *musts* for the vehicle include a white, four-door sedan with leather interior, GPS, a minimum of 30 mpg, and an iPhone-compatible sound system. Your *wants* include power windows and seats, all-wheel drive, voice-activated controls, and six cup holders.

At this point, all other vehicle ideas that fall outside the musts list such as sport utility vehicles, vehicles getting below 30 mpg, non-white vehicles, etc. are immediately thrown out as being nonviable (or bad ideas, if you wish). From all remaining possible vehicles, it becomes a value-sorting process from among those with varying degrees of built-in wants.

From this value-sorting process, you have identified the best idea for a vehicle based on your own personal perception related to identified qualities—accessories, features, and other points of comparison—that you have determined as essential for an idea to be considered good. If your musts included criteria such as ruggedness because you work in construction, or luxury because you wish to portray a certain image, then you would have started the idea evaluation from a very different point. For most, this isn't always a very conscious process and happens somewhat naturally.

Most definitely, though, the decision of what's good is based on personal judgment that, aside from the measureable qualities discussed earlier, is also determined by a number of other secondary attributes based on bias, desire, emotions, feelings, education, taste, and experience (such as the vehicle's shade of white). These secondary attributes are what create the perception of good: a sense of "attractiveness," "excellence," "superiority," and "worthiness." These attributes are very difficult to observe, count, or quantify, and much easier to measure with a gut feeling. The value or goodness of something comes as a result of the desirability, purpose, and use that we as people place on things that satisfy our musts and wants and are often the result of choice.

Outside In

When choosing from a variety of ideas, an honest assessment and understanding of our personal bias will help to clarify the judgments and decisions made during the imagineering process. This honesty will allow us to keep an open mind while systematically evaluating the development, selection, and implementation of ideas against the other possibilities that exist.

One method is a conscious, value-sorting approach that requires an "outside in" process to evaluating ideas. Since we are going to distinguish between good and bad based on our own personal criteria (or group criteria), it's easiest to begin with the extremes since they tend to be the easiest to "see" based on perception and gut feeling. In other words, when ranking ideas or solutions from a list of possibilities, begin by picking the best and worst from the group and placing them on the outer ends of the "goodness" continuum that ranks ideas from best to worst (3 and -3 in the diagram below - see Figure 6-1). Repeat the process and place the next best and worst from the group on the continuum (2 and -2). Regardless of how many possibilities you are ranking in terms of goodness, continue this process until you reach the last ideas from the list, which are placed in the middle. You will have successfully distributed the ideas from "good" to "bad" based on your personal perception and criteria. If a group of people is performing the same process for the same list of options, lists can then be combined to create a group scale that could show some consensus. Although the diagram below only accounts for seven different ideas, this can be done for any total number of ideas.[6]

Of the good ideas, how efficient are they? What is the likelihood of success for each? This is when you perform some mental elaboration on the abstract ideas to further refine the selection. Although the highest rated idea will best fit within your or your group's primary and secondary attributes, it may or may not be the best overall idea after further discussion and analysis. If an idea is chosen and implemented, it's imperative to keep a close watch to make sure the idea achieves desired benchmarks, as only time can show the wisdom or folly of choice (like having gone to the Moon).

Of course, aside from our personal perceptions related to a variety of characteristics, there are a couple of other influencers that can directly impact how we view whether or not ideas are good.

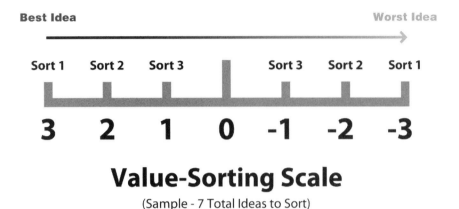

Value-Sorting Scale

(Sample - 7 Total Ideas to Sort)

Figure 6-1.

Attack of the Killer Kirby

I remember the first vacuum cleaner I purchased back in the late 1980s—an eight pound Oreck upright. It was compact, lightweight, and easy to use. Based on my income at the time, it wasn't cheap. That vacuum worked great, and I owned it for many years. Then one day in 2001, there was a knock at my door. It was a Kirby salesman. In a world with the Internet, it was hard to believe that door-to-door salespeople still existed; yet there he was, vacuum in hand. At first, I thought it was being delivered to my house by mistake. But, after some smooth talking on his part, there he stood in my living room, dumping a full cup of sand all over my carpet.

Before I even had a chance to complain, he asked if he could use my current vacuum to clean it up. I proceeded to get it out of the closet, and he started vacuuming until we both seemed convinced that my trusted Oreck had done its job. He then whipped out the Kirby (with a special clear canister instead of a bag so we could clearly see what was sucked up) and vacuumed the same spot again. I couldn't believe what I was seeing; the Kirby had sucked up a ton of sand my vacuum had missed. And, like the sand in the clear container, I was sucked right into buying the Kirby on the spot (and it proceeded to vacuum out my wallet as it was considerably more expensive than the Oreck I'd bought years earlier). Cost wasn't an issue though—especially in light of what I had just witnessed.

The Suction of Vacuums

I believe this story illustrates a common problem today. With the rapid pace of daily life (which only seems to be accelerating), people often get caught up in what I call "attention vacuums." As an avid observer, I have noticed a growing tendency in which people allow the "stuff" in their daily lives to suck them in like giant vacuums. A vacuum, by definition, is the absence of matter. Nature abhors vacuums and does whatever it can to fill the empty void with whatever immediately surrounds it. A personal vacuum occurs when one's thought process becomes overly focused on a specific situation or event (like dumping sand onto my carpet). Our attention becomes fixated on and sucked into the here-and-now and the specific situation we are dealing with at the time, thereby causing us to miss the bigger picture.

We become overwhelmed by the many "details" of a specific situation and still try to balance everything with the hope of being successful and happy. Our energy gets divided into small bits, spread over a large number of simultaneous actions. Individual activities seem to have a higher priority than the holistic nature of them, and as a result, people often find it more difficult to make everything work together toward one common purpose for the long-term. The suction of vacuums, a huge purple people eater, can have a direct, negative impact on our ability to visualize anything new, better, or different, and can negatively alter how we perceive and assess the goodness of ideas at any given time.

For example, think back over your life. Can you remember a past event which, at the time, seemed so overwhelming or troublesome that you had no idea how you were going to deal with it, let alone survive it? How about that breakup with your first love or when you were laid off from your job?

Throughout the latter part of my high school years, I dated someone I would define as the "first love" of my life. She was a girl who, as far as I was concerned, could do no wrong. However, my parents didn't quite see her the same way, and my mother often referred to her as "poison" because (in my mother's view at least) she was poison to my ability to make rational, mature decisions. Of course, my mother's perception only entrenched my feelings toward her. When we eventually broke up after a few years, I was devastated; it affected every aspect of my life. I didn't feel like eating, leaving the house, or doing anything, for that matter. I remember sitting in my bedroom for hours listening to the Styx song *Babe* over and over. It was the end of life as I knew it.

Eventually, though, I got over it. I later realized that not only was my mother right (I had indeed made some very poor decisions), I had allowed my relationship and the subsequent break-up to divert me well off the course I had set for my life. Even worse, I had no idea I had gotten off-track until it was too late. "Poison" had become a vacuum in my life that sucked all of my energy, diverted my attention, and caused me to incorrectly evaluate some ideas.

Vacuum events often seem like the black holes of the universe. They require mass amounts of energy and pull our focus away from the long-term to everything short-term and immediate. In hindsight, they are seldom as big of a deal as we made them out to be, especially in the context of life's big picture. More often than not, we are ultimately able to survive and move past them, and we probably become better people as a result.

Because the distractions in our lives seem to be increasing at an exponential rate, I see vacuums having an ever-growing impact. People spend so much time dealing with such a large number of little vacuums, they become bogged down like a car spinning its wheels in the mud. When living in a world of vacuums, we become overly focused on specific details without stopping to contemplate the potential long-term impact. We get sucked into believing that every-thing is equally important, which distorts reasoning and keeps us from seeing how specific situations or problems fit into the overall scheme of things.

Vacuums are the archenemy of productivity and goal-directed behavior. Vacuums work overtime to divert the thought process down an unproductive mental tunnel and, if allowed, they can choke the process of assessing ideas like weeds in a garden. The effective thinker is very conscious of the negative effects caused by vacuums and fights hard to resist the sucking power of distraction.

The Suction Blocker

Throughout life we are confronted by countless problems to solve and decisions to make. Some are extremely important while most are relatively insignificant and fruitless. Some will require a rapid decision while others pose no real time constraints. Some carry a high level of risk while others are risk-free. Some are complex while others are fairly simple and effortless.

However, regardless of the problem to solve or decision to make, the only way to effectively determine which ideas are truly good (and avoid the vacuum effect) is to maintain a conscious, "tan-gible" sense of direction. This tangible direction allows you to create and maintain daily perspective, thus providing you the power to resist and fight off the continuous suction of vacuums we all face.

Allow me to illustrate. One thing is true in terms of human behavior: over the long-term, behavior will usually align with a person's personal values. In other words, although people will routinely compromise their personal values in the short-term for a variety of reasons, eventually their behavior will realign toward what they deem to be most important in their lives. In my case, it was life itself.

A number of years ago, I had my routine annual physical. Everything was fine except for a bad liver score on my blood test. Under "normal" circumstances, my physician said he wouldn't have been too concerned about it since a couple of mixed drinks consumed shortly before my blood was drawn could have caused a similar spike in the test. One problem, however: I didn't drink.

To be safe, he ordered an ultrasound of my liver, and the test confirmed what he had suspected. I had non-alcoholic fatty liver disease.

He told me that the condition was more than likely the result of bad genes, not anything I had necessarily done, like overeating. Please understand, I wasn't "fat" in the traditional sense, but I also didn't give a second thought to what I put into my mouth. I was so busy living each day that, like most people, I ate fast food frequently and didn't exercise very much. Over time with age, my weight slowly crept up to the point where I could easily stand to lose a few pounds, like maybe five or so (at least that's what I thought).

During my next visit, my physician shared both the facts of the disease and the only known effective treatment according to most studies; I had to lose 15% of my body weight in a short period of time. In other words, he wanted me to lose about 35 pounds in less than six months!

In order to achieve such a daunting goal, I knew I needed a tangible vision of what the future looked like if I wanted any chance of making changes to my daily behavior. I knew I wasn't going to lose the weight overnight, especially since I spent years putting it on. I needed something I could strive for—something I could put in front of me at all times to remind me why I had to stay away from fast food, sweets, junk food (basically anything that tasted good), and why I would be sweating so much on my treadmill.

I could have easily picked a photo of some young "ripped" guy like those featured in exercise equipment commercials, but I knew that wasn't realistic for me. I needed real.

I picked three photos: one of a fatty liver, one of Jared (a 450-pound college student before he started eating Subway) and one of Jared after he dropped a couple hundred pounds doing it.

Jared's father, a physician, had used fear as a tactic and basically told him that if he didn't lose weight, he wouldn't live past 35. Jared was so large he chose his course schedule in college based not on the courses he wished to take or who was teaching them, but instead on which rooms had tables that would facilitate his enormous body. He started eating a foot-long veggie sub at lunch and a six-inch turkey sub at night. He lost 100 pounds in three months. After that he began to walk more and the weight started coming off even more quickly.[7] (*Note:* I chose Jared as my inspiration many years before he got sucked in by some very poor vacuums and ultimately went to prison for it.)

Each day I looked at all three photos to remind me of what I was trying to achieve and to help keep me focused on the purpose behind all of my "suffering." I also studied all of the current diets that were available, and after reading the summary of a Harvard University long-term study on diets, I decided my best course was to just eat less and exercise. I committed myself to the idea of eating no more than 2,000 calories per day with regular weekly exercise until I lost the weight, and then increase my daily calorie intake to

2,500 once I reached my goal. I kept a journal of both calories burned through exercise and the calorie count of everything I ate to ensure I was staying within my daily goal (and to help me resist the countless vacuums that regularly appeared, like the cookies brought into the office or the cheesecake sitting in front of me at one of the many event dinners I attended).

The result: I lost 45 pounds in four months. I have since stopped keeping written track of ingested and burned calories. However, having done that for a few months, I managed to memorize most calorie amounts in foods and to this day keep a running total in my head in order to consume about 2,500 calories or less daily. I have been able to keep off "most" of the weight because the short-term changes to my daily behavior patterns resulted in relatively permanent changes in my lifestyle. As a result, I am able to resist getting sucked in by food vacuums, although I admit I occasionally cheat now and then, especially on holidays.

Granted, to a certain degree I was motivated by fear since my physician basically said lose the weight or die, but I was also motivated by the value I had for my life. Whether it's life, family, employees, finances, God, or country, understanding that which we value helps us create a tangible personal vision. This results in a greater ability to take control of daily activity, properly evaluate ideas, and achieve long-term success.

Ultimately, the secret to fighting off vacuums and achieving what you want, regardless of your situation in life, comes down to two things:

- Keep a tangible picture of what you want in front of you at all times to serve as a constant reminder.

- Make good, daily choices that reflect the values represented by that picture.

Like most things in life, success is achieved by the DAILY decisions we make over time, not the decisions made in a single day.

Haste Makes Waste

When I was a little boy, go-karts were very popular in my neighborhood, and many of the kids had great-looking karts—their dads had clearly spent a bunch of time constructing them. Obviously, I too wanted a kart I could push around and ride down the hill near our house. The road on the hill was paved with tar and then covered with pea-sized gravel. This "paving process" comes into play later.

When I asked my dad to help me build a kart, he took one large board, slapped a couple of smaller boards to the bottom of it in order to hold the axles and wheels (the front board was able to pivot), and placed a rope on both ends of the front board to be used for steering (see Figure 5-2 below).

Needless to say, Dad wasn't too worried about visual aesthetics and was far more concerned about getting me started with my "new ride" as soon as possible. However, there was one major flaw in Dad's rapid design; he used completely threaded steel axles that extended a few inches beyond each end of the smaller boards with large threaded nuts to hold on the wheels. It seemed like a good idea at the time and an easy way to attach the wheels. However, each time I took the kart to the top of the hill and let gravity do its job, the spinning wheels would gradually unscrew the nuts from the axles and one or more of the wheels would come off while the kart was in motion. You can picture the rest.

My dad is one of the smartest people I know, and an exceptional craftsman when it comes to building furniture and doing many other hands-on projects (ironically, if he built a kart for me today it would probably look like a Formula One Indy Car made of wood). Granted,

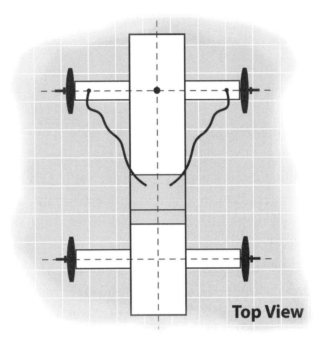

Top View

Figure 6-2. A top view of my father's go-kart design.

the kart construction occurred over 45 years ago before he had fully developed his skills, but this isn't a story about skill. It's a story about haste and failing to completely think through the long-term outcomes of very short-term action.

Throughout my entire childhood, my mom used the phrase "Haste makes waste!" to emphasize what would happen if I rushed something or hurried through tasks rather than doing them correctly. Her intention was to get my brother, sister, and me to slow down, take our time, and do our best job, no matter what we were doing.

I now realize how almost comical our lives have become today as we try to cram as much activity as possible into a finite amount of time. We justify this "busyness" and claim we're more productive, but the result is a life filled with added undue stress, a growing lack of long-term focus, and a decrease in the effectiveness related to evaluating ideas. I regularly see people working frantically each day, performing tons of tasks and activities, expending a bunch of energy, and yet they seem to accomplish very little of significance before becoming frustrated.

Want Fries with That?

People are so busy being busy that we have nearly created a total "take-out" existence. Our expectation is to do ever more while achieving instant response and gratification. With changes in technology, delivery systems, and social norms, we now live in an immediate, take-out culture:

- Automatic teller machines and real-time online banking

- Pay-at-the-pump gas

- Instant credit approvals for purchases (that you may or may not be able to afford)

- Pills to rapidly fix a host of physical problems or psychological issues

- Fast food with drive-up windows or deliveries that can arrive in only 30 minutes or less

- Even more drive-up windows in pharmacies, dry cleaners, liquor stores, banks, coffee shops, bakeries, and grocery stores

- Same day and overnight delivery services

- Email, voicemail, text messaging, tweeting, and the smartphones that facilitate it all

- "Hanging out" with friends on social media sites and blogging about all of it from your sofa

- Online and downloadable movies, music, television, and virtually all other forms of entertainment without leaving your house (or sofa)
- Expedited checkout and ordering through barcode scanning
- Web-based education
- 10 minute oil changes
- Virtual reality
- Schemes for rapid weight loss, quick wealth accumulation, and fast exercise routines
- Instant bill pay
- Complete microwavable meals made in seconds
- Speed dating events and online dating services

And the list of life-accelerating examples goes on and on. With the growing rate of "now" in our lives, one thing is becoming increasingly clear: we now live in a society where flexibility, immediacy, and instant gratification have become the norms in life, so much so that I've come to think of the youth of today as the "generation of instant", or iGen (to borrow the "i" from Apple).

Notice those around you. People seem to have very little patience, complain when they don't have enough time to do everything, have shorter attention spans and a lack of focus, and expect change or things to happen quickly. As a result, these same people don't seem willing to allow ideas to play out until the end, to make decisions that may require considerable time to accomplish, or to focus long enough to complete a goal. All are potential purple people eaters, and at the very least, as my mother would say, it's haste making waste.

Having taught both undergraduate and graduate college courses and having held a variety of leadership positions over the years, I've witnessed a growing discontent and frustration by people concerning the lack of time they have to complete tasks. But, time moves the same for everyone. What it really comes down to are priorities, choices, and vacuums. People ultimately choose what they will do and prioritize the order in which it will be done, yet they still allow vacuums to alter how they evaluate ideas. (Remember, a personal vacuum occurs when one's thought process becomes overly focused on a specific situation or event.)

Personally, I'm no different from anyone else. My life sometimes seems to blur together without real distinctions between work, home, and play. I get frustrated by flight delays, Internet delays, slow computers, slow drivers, slow customer service, and pretty much anything else that slows me down. However, I have made considerable progress in the area of imagineering and decision-making. Because of many poor past decisions, I have now become very aware of my time and am more proactive with how I use it.

In order to be effective in the long-term, I have had to take control of my short-term DAILY activity. I now force myself to stop and ask, "Why am I doing this and what purpose does it serve?" I have learned to:

- Become very conscious of how I use my time as it is very precious and limited;

- Become very conscious of vacuums and how they, more than anything else, can continually pull me away from choosing ideas wisely;

- Minimize the extraneous "stuff" in my life, especially those activities that don't help me reach my ultimate goals;

- Work at a slower, more thoughtful pace, which actually makes me more effective in assessing whether ideas are good or not.

All of these decisions have improved the quality of my work.

Impact on Leaders

Our take-out, instant gratification culture has also complicated the role of leaders. Leadership, personal or otherwise, is all about the quality of decision-making, and one of the biggest decisions leaders face is long-term success versus short-term gain. As more leaders are now being evaluated on shorter-term outcomes, the temptation is to make decisions that reflect greater short-term gains, regardless of how those decisions may or may not affect the long-term.

In other words, to steal a phrase I once heard from a friend and college basketball coach, "You're only as good as your last season." Since people have less patience, we as leaders are asked to show rapid, short-term gains that once took much longer to accomplish. Whether you're a coach, politician, CEO, president, sales manager, church pastor, or a parent, a growing pressure exists to exceed short-term expectations, unrealistic or otherwise, in whatever manner possible. The result is often rapid success or positive outcomes, followed by a downturn of some manner in the following years.

For example, in 1997, Jim Leyland, a very talented baseball coach, was hired by owner Wayne Huizenga to coach the Florida Marlins, a team only in its fifth year of existence. With a number of very talented players ("purchased" for a total of $89 million, one of the highest team payrolls in the league at the time), Leyland was able to coach the Marlins to the World Series and the expansion franchise's first championship. Immediately following that great season, Huizenga claimed he was losing millions of dollars and immediately began to dismantle the team in what became known as the "great fire sale." As a result, Leyland had a terrible 1998 season and resigned soon thereafter.[8] This example illustrates that with enough short-term effort or investment, great short-term outcomes can be achieved. However, without a solid basis or foundation built on a strategic focus, sustaining positive outcomes is difficult, if not impossible.

When Haste Doesn't Make Waste

Up to this point, I have primarily focused on being effective and actually choosing good ideas for the long-term. However, efficiency is just as important, and there are times when taking a short cut or accelerating a process can actually yield long-term positive results. While effectiveness can be viewed as "choosing the right ideas," efficiency can be viewed as "minimizing the amount of required resources to execute them."

Like the phrase that Nike coined in 1988, sometimes the best strategy is to "Just Do It"—just jump in and get the job done because that's what needs to happen. Yet, many people delay making decisions because of the risk associated with uncertainty. As long as this short-term process has been thought through in the context of a long-term strategy (and this is key), decision-making can yield a greater level of efficiency by using less time, money, and other valuable resources to value-sort ideas. We are able to accelerate the process of accomplishing desired long-term goals and objectives through more rapid, yet effective, short-term decision-making.

When a child is taught how to ride a bicycle, she usually gets a bike with a set of training wheels to support the bike during the learning process (or at the very least, the child is placed on the bike while a parent runs along the side holding it steady). Over time, the parent will gradually inch up the training wheels until the point where the child is no longer using them. This is typically followed by a grand ceremony of taking off the training wheels in front of the child as if she had just "graduated."

For whatever reason, my dad decided that an expedited, more direct approach would be better. On that same freshly tarred-and-graveled hill where I rode my go-kart, my dad took me to the top, placed me on the bike, and let me go. Of course, I didn't stay on it very long and fell. Dad just picked me up, brushed off the sticky gravel, placed me back on the bike, and let me go again. We repeated this process until I was riding on my own. Needless to say, I caught on quickly. I either had to learn how to ride or endure the pain of repeatedly falling and becoming "tarred and graveled."

I'm not sure how much time Dad spent thinking through this "teaching moment" (which some people today might find somewhat harsh), but his actions had a huge impact on how I approach and deal with goals, tasks, and obstacles. Instead of slowly and incrementally approaching them, I have a tendency to quickly size-up a situation, make a decision, and jump on the bike, so to speak. But, unlike speed for the sake of speed, he taught me the most efficient path to success can be the quickest, but also the hardest, because it requires determination and a willingness to endure short-term pain for long-term gain.

Of course, I've made mistakes and poor choices, just like everyone else. However, I have come to realize this one simple teaching moment actually had less to do with learning to ride a bike and far more to do with providing me a greater lesson, as I now possess the tools and drive to be more proactive and determined to succeed at whatever I'm doing.

Whether intentional or not, thanks for the lesson, Dad. And thanks for the go-kart too. A simple kart, even one with a flawed design, was much better than having no kart at all.

And the Final Point...Practice Kaizen

Selecting what you believe to be the best idea isn't the end of the story. Ideas aren't static; they require continuous imagineering. The Japanese call this *kaizen* (pronounced ky'zen), the relentless quest for better. Think of it as the daily pursuit of perfection. *Kaizen* keeps one reaching and stretching to outdo yesterday. Although the improvements may only come a little at a time, these small, incremental enhancements will eventually add up to become something greater.

Continuously improving ideas requires making new connections between current ideas and possible new opportunities. For example, Fred Smith began FedEx in 1971, long before the advent of the personal computer, Internet, and mobile computing. But, these ideas and many like them have been added over the years to his original idea of overnight shipping, thus enhancing the shipping process through greater effectiveness and efficiencies in tracking and logistics. As author Tom Peters (best known for his book, *In Search of Excellence*) once put it, "If we're not getting more, better, faster, than [our competitors] are getting more, better, faster, then we're getting less better or more worse."

One method of practicing *kaizen* is to not get lazy when it comes to learning. In our fast-changing world, it doesn't take long for skills and knowledge to quickly become outdated. Advances in technology and the flood of new information have made it hard to keep up. Lifelong learning is the only way to expand your box and remain truly creative. The more you know, the easier it is to make new and ongoing connections, continue to imagineer better ideas, and keep the big idea as the big idea.

Taking a Few More Steps:

• What does the word "good" mean to you? Take a moment and write your personal definition of it. How does that definition affect how you view outcomes?

• Create a "sounding board" (1-3 people you totally trust) to express your feelings, bounce ideas, vent, whatever. Be open to their feedback, no matter how brutally honest it may be.

• What shapes your current "box"? What obstacles (vacuums)—such as fear, insecurity, financial constraints, or some other self-imposed limitation—get in the way of expanding it? Look deeper and be honest with yourself. Otherwise you can't fix whatever holds you back and face obstacles head-on.

• Feel like you're on the sidelines suffering from "spectatoritis"? Get in the game and do something that helps you fire up a little passion. People "buy into" feelings, and passion gains support for ideas.

• What do you do well? Do it. What do you enjoy doing most? Do it. If they are not the same thing, then find a way to merge the two.

Not My Monkeys, Not My Circus

If you're like most people, odds are you're swamped—so much to do, so little time to do it. We wade through our days trying to balance ever-growing responsibilities, and when we do them simultaneously, we feel more productive. We call this *multitasking*, and we believe the better we are at it, the more effective and efficient we will be. We tend to view multitasking as a positive, frequently sought-after attribute. In fact, as many of you read this, you're likely responding to text messages, checking emails, eating lunch, reacting to app notifications, and thinking about the rest of your day at the same time.

Multitasking is a myth. Sure, you can chew gum while walking, listen to music while vacuuming, eat lunch while reading, or fold laundry while talking on the phone. But, these activities don't require higher-order, problem-solving skills or much brainpower of any kind. Psychologists who have long studied the concept of multitasking have found that the brain is unable to focus on more than one higher-order function at a time. When people multitask, they actually shift their attention from one thing to another at fast speeds, and each time they switch focus between tasks, their minds must cope with the new information.

What is actually occurring is *switchtasking*. According to Earl Miller, a professor of neuroscience at MIT,

> *"People can't multitask very well, and when people say they can, they're deluding themselves...Switching from task to task, you think you're actually paying attention to everything around you at the same time. But you're actually not."*

There are several reasons for this, but one is that similar tasks compete to use the same parts of the brain. For example, talking on the phone and writing an email are nearly impossible to do concurrently because of what neuroscientists call "interference." Both tasks involve communication skills and contend for similar space in the brain. Multitasking doesn't actually make us more productive; in fact, the quality of our effort suffers. Another major downside to multitasking is the negative effect it has on our stress levels as we try to balance a multitude of simultaneous activity. As a result, we feel overwhelmed, drained, and anxious.[1]

Why has multitasking become so important? It hasn't always been this way. I can remember a time not too long ago when people were pretty content doing just one thing at a time and living much slower lives. What's different? I believe the answers can be found by looking at two distinct yet interrelated aspects of everyday life: *technology* and our level of *happiness*.

Technology

When I was in high school, personal computers were barely in their infancy. Way too expensive for the vast majority, PCs with any real productive power were only found at the corporate level. Some high schools and colleges were beginning to use them, but for the most part, the average person still had little to no personal contact with a computer. Cell phones didn't exist, let alone anything remotely resembling today's power-packed smartphones.

In other words, by today's standards, people were pretty disconnected. To communicate, you either made a call from a bulky telephone connected to a wall, talked face-to-face, sent letters, or fired up your CB radio (if you were born after 1980 you may need to ask someone older about this).

The lack of accessible personal technology resulted in a slower life, one that required more planning and coordination to maximize productivity, stronger interpersonal skills, and greater levels of patience.

Current technology demands an entirely new context: one where people spend less time planning their days since most things can now be done on the fly; one where the need for interpersonal skills between people continues to diminish as a larger percentage of our communication is now virtual; and as discussed in the last chapter, one where expectations of "instant" are now the norm. Be honest, after you send a text message or leave a voicemail, how long are you willing to wait for a response before feeling frustrated...even a little?

This change in thinking—especially for younger generations who only know this type of thinking—combined with the ubiquity of personal electronics has resulted in daily expectations of immediacy and convenience. Ultimately, we feel like we're doing more in less time, and thus create and perpetuate the concept of multitasking.

Unfortunately, while technology has definitely become more capable, our minds still basically work the same. The result of this ongoing pursuit to do more in less time is ultimately the diminished quality of our efforts with increased levels of stress and anxiety.

Happiness

As humans, we want to be happy. Most of us believe it's a basic human right, and thus we have a tendency to try to use whatever we have at our disposal to acquire it. Whether it's through status, stuff, or other people, we have a desire to feel valuable in our own eyes and in the eyes of others.

The challenge is to know what it means to be happy. Although I know a number of people who think happiness is complicated and dependent upon a large number of factors, I tend to believe happiness is nothing more than a function of both *expectation* and *reality*—the relationship of two independent variables that ultimately affect our feelings of happiness. A mathematician, or any one of the dozen or so people in the world who aren't afraid of math, might view it like this:

Happiness = f (Expectation, Reality)

As long as someone's reality—perceived or otherwise—is above their level of expectation, they generally tend to be happy. However, when those pesky expectations start rising too high or even stay the same while our current state of reality declines, unhappiness typically sets in.

I believe that as average people we have the most control over, or can more directly impact, our levels of expectation. Life's outcomes and subsequent realities are typically not in our direct control, since rewards and other positive changes are frequently at the behest of others. So, let's focus on a few aspects in life today that can directly impact our expectations:

Hedonic Adaptation: *"Hedonic adaptation" is a psychologist's* *way of saying the novelty wears off.* Eventually, that new house, car, or smart TV you had to have becomes just another thing you own, or the job you worked so hard to get becomes just part of your daily grind. *Your lifestyle adapts, and you're back to wanting more.*[2] I've now come to embrace that happiness related to "stuff" is a choice, and there's nothing tangible that can "make" me happy in the long-term. No matter what we work toward or feel like we must have, typically the happiness attached to it is only short-term. Each time you receive that "must-have" thing, it only serves to raise the bar of expectation for the next must-have thing.

Social Media Image Crafting: We are always trying to put our best foot forward and want to look good to others. It's human nature. With the advent of social media, you can take it to an entirely new level and present yourself any way you wish, and it's usually positive. According to a piece on Whole9life.com (a health and quality of life website), *"Our social media feeds read like a modern-day fairy tale, where every moment is wondrous, every interaction with our family is more precious than the last, and even the mundane (Coffee with the girls! Look at my lunch! Stuck in traffic!) is a magical experience."*[3] Social media image crafting tells everyone that a perfect life is not only attainable, it's normal. When everything about your social media "friends" seems perfect, it naturally raises the bar of expectation related to your own "imperfect" real life; thus, the gap between expectation and reality is potentially widened causing increased levels of unhappiness.

Technological Overdependence: Frequently, happiness is thought to be the natural result of success. Although an extremely subjective term, "success" for many of us often revolves around the feeling of being busy, as "busyness" implies productivity. As mentioned earlier, technology helps to provide this feeling of being busy. Naturally, we expect our technology to always work the way in which it was designed. When it doesn't, it causes stress and anxiety. I was once at a busy grocery store when their computer network suddenly went down. Check-out registers could no longer take debit or credit cards. People had to use cash. Since most people today typically don't carry much cash, there was a mad scramble to the ATM, which was quickly emptied (it was on its own, separate network). Chaos, anger, arguing, yelling, and frustration all ensued. Much unhappiness was present. By the way, I did happen to have cash, so I got to watch and be entertained—and a little scared—by it all.

Future-Focused: Too often, we overly set our sights on the future, and we can only see the present after it has become the past. Being goal-driven isn't a bad thing, unless we are too future-focused, and then our expectations of future joy can blind us to the joys and value found in the now. We may frequently find ourselves absent from the moment as any one of a great number of distractions pulls our attention in a variety of directions, all with the intent of getting or achieving something else "down the road." If getting older has taught me anything, it's that time is finite. There's never enough. I'm amazed at the growing frequency of what I call "time-lapse realizations" that occur the moment I accomplish some goal or objective. While I'm happy I achieved what I set out to do, a sudden realization often follows: getting there came at a great price. A feeling of emptiness often overtakes me, as if I had been transported into the future with little memory of the daily joys from the actual act of doing. I realize how fast time raced by, and because I was so goal-oriented, I was unable to fully enjoy the experiences related to the process.

Childhood Letdown: My good friend and author Adam Carroll frequently talks about how we as parents can sometimes love our children too much. It happens in a variety of ways: giving them things they should have had to work for, not helping them to understand the true value of something, or by setting high expectations for them that are impractical once they become self-sustaining adults. Sometimes, in our efforts to "encourage" or "inspire" them to become successful or achieve greatness, we provide motivational but unrealistic guidance. How many parents have told their children that they can be or do "anything" they want when they grow up? According to the *Book of Odds*, the probability of becoming the President of the United States is 10 million to one. The probability of becoming an astronaut is even greater (believe it or not)—12.5 million to one.[4] The unfortunate, negative side effect to all of this is the potential of setting children up for failure and disappointment because expectations were set too high.

When I was a child, I was presented with some amazing opportunities. I was able to represent the United States in the International Science and Engineering fair and worked in the research and design department of a computer manufacturer, all while still in high school. Needless to say, many in my family were convinced I would become the family's first multi-millionaire; a view they *often* shared with me. I'm now in my fifties and am still working on that millionaire thing. Not to say that I haven't been successful in life, but those words still haunt me a bit today, making me question, "What could or should have been?" and "How have I possibly fallen short of my potential?"

Happiness is a state of mind impacted by where we set our expectations. While these five aspects and many other factors directly affect those expectations, we are ultimately in control of where they're set in relationship to our current state of reality.

While "strategic" and long-term goals are definitely not bad in and of themselves, they will seldom ever be achieved if set at levels requiring too much time to realize. The gap between reality and expectation will be too great and, ultimately, results in unhappiness. Had Kennedy set the goal of going to the Moon by the end of the century (39 years) instead of the end of the decade (9 years), I believe we would have never made it. The increased amount of time would have allowed too many other variables to enter the equation such as changes in leadership, government, social issues, and a host of other aspects that would have eroded the drive and intensity necessary to reach the Moon.

Think tiny. Ideally, our expectation bars should be set at short, attainable levels so both growth and happiness are incremental. Small, short-term accomplishments will not only serve as a motivator toward the future, they will help you maintain an achievable level of ongoing happiness. After all, isn't that what everyone wants?

The Art of Singletasking

Although I can't claim to be a model of efficiency (I too, at times, get sucked into the false hope of multitasking), I have learned over time how to effectively maximize my efforts and accomplish large tasks and projects. Time is finite, and we all have the same amount of it—24 hours per day/7 days per week. In order to maximize my productivity, I've adopted six strategies that greatly impact not only the number of positive outcomes I can accomplish, but also the quality of my work.

1. Be Like Syd

I own a beautiful liver and white springer spaniel named Sydney.

Sydney does four basic things in life and never at the same time: eat, play, poop, and sleep. You can't ask for a more simple life, and despite that, she's happy. And, she's always present in the moment.

We need to be more like Sydney and simplify our lives and stop trying to do everything simultaneously. Our brains are complex organs. The average human brain uses the equivalent of 20 watts of power (enough to power a light bulb), and although the brain only comprises a mere 2% of our total body weight, it consumes more than 20% of our daily caloric intake—more than any other organ in the human body.[5] Research has shown that our mental energy related to decision-making is finite, and once depleted, the quality of our thinking begins to dramatically suffer. As average people, we tend to spend a large percentage of our mental energy on relatively meaningless stuff that really doesn't have any real impact on our lives, good or bad, like streaming through countless posts on Facebook and watching television. Once our brain has used its energy, we tend to miss the relevant stuff and other important details necessary to be more successful, creative thinkers within the limited time we are given.[6]

Studies of very efficient people show they rid themselves of distractions and the unnecessary, miscellaneous choices that deplete mental energy. They frequently eat and meet at the same places; they turn off their smartphone app notifications and look at their apps when they're ready to see them; they stop dwelling on things that occurred in the past and don't obsess on things that might happen (since it's impossible to actually do things in the past or future); they frequently wear the same clothes (think Steve Jobs); and they remove the clutter that surrounds them.[7]

To illustrate the power of simplification, consider the high school equivalency GED exam, which has been around for over 70 years. Recently, the exam shifted from paper to a computerized format. Unlike the paper version, where multiple questions along with multiple answer slots were all visible at once, the computerized version removed the clutter and only showed one question at a time. The passing rate on the computer exam rose to 88%, compared with 71% for the paper version (a 17% increase).[8]

Being at our creative best requires gas in the mental tank, gas that will only be available if we aren't going full throttle every day. Be like Sydney. Simplify your life.

2. Write to Think

Studies have shown that writing, pen or pencil on paper, slows down your brain and actively engages your senses.[9] Writing typically generates more "meaning" in the mind than does either tapping or thumbing a keyboard. When we talk, we typically don't think about the words we are speaking—hence, the proverbial foot-in-mouth. When writing, we do.

Writing puts your thoughts more in-line with your feelings which helps to convey greater personality similar to how an artist feels when putting paint on a canvas. When you are using a pen (or pencil), you tend to feel more artistic which helps bring out greater levels of creativity.

As an added bonus, writing on paper will never "crash," get accidently erased, get inadvertently spread throughout the Internet like a virus to come back to haunt you someday, or require power when your battery dies.

3. P.E.E. Daily

It's human nature to deal with things as they occur. Our days have a tendency to fill on their own; as a result, we find ourselves regularly "putting out fires," thus becoming continually distracted by the constant change in events. With full lives, we can create a pattern of putting off what we "want to do" in order to deal with what we feel we "must do." After talking with so many people over the years who have been unable to realize their goals or dreams, I've come to believe that the saddest word in the English language is the word *someday*, as in someday I'll write a book, someday I'll get out of debt, or someday I'll run a marathon.

The only way to do anything of significance and do it well— and thus achieve those goals and dreams—is to Plan, Execute, and Evaluate (P.E.E.) on a daily basis. This process is all about the **_DAILY_** discipline of *planning* how to use your time, *executing* by taking a purposeful step forward to accomplish a desired result, and then *evaluating* it to make sure it was actually a good step since not all daily progress is actual improvement.

To P.E.E. is all about daily discipline. I spend the first 10 minutes each day planning that day and the last 10 minutes relaxing in bed reflecting on it and planning the next day before falling asleep. Those 20 minutes I spend each day not only serve to help maximize my daily productivity, they make a huge difference in helping me to successfully accomplish my goals.

To P.E.E. is also about proactively scheduling time—even if in short amounts—when you will work on ONLY one task and nothing else. To write this chapter, I had to schedule two days out of town when I would do NOTHING except write. I blocked out six hours each day. I made sure my dog was cared for, I took care of my physical needs, and I placed a cool beverage and snacks by my side prior to beginning. During those twelve hours, I did nothing else. I have employed this strategy for virtually every task of significance whether writing a book, designing a project, preparing a presentation, or cleaning out the garage. Not only am I able to accomplish each in its entirety, I'm also certain the quality of my work significantly improved because I remained focused.

4. Unplug

Like an addict goes to rehab to cut off access to addictions, being productive requires the same. Even if you set aside time, it won't work unless you unplug and disconnect. Otherwise, distractions will magically follow. The phone will ring, text messages will appear, emails will arrive, and you will receive a multitude of notifications from Facebook, LinkedIn, SportsCenter, Words with Friends, and all of the other apps on the smartphone holstered to your hip, all acting as huge temptations screaming for attention. Like most addicts, we will be unable to resist these distractions. Turn off the smartphone. Disconnect the laptop from the Internet. Short of an emergency or crisis (which seldom ever happen despite worries to the contrary), everything can wait. Although there's the possibility you may feel some form of separation anxiety, it's absolutely necessary in order to focus. Even Bill Gates, during the hectic growth period of Microsoft in the 1980s and 1990s, realized this. Each year, he would set aside time to relax, disconnect, and use the time alone to read and think.[10]

5. Use Five-Star Locations

Imagine getting out of a cab at the entrance of a five-star hotel. You immediately notice the smell of fresh-cut grass, the beautiful landscaping, and flowerbeds. As you enter the lobby, you can feel the elegance of the décor. The staff is dressed quite handsomely. The property is very well maintained. You hear enticing music and laughter from the lounge. The smell of hot chocolate chip cookies at the front desk masks the slight smell of chlorination from the nearby fountain while you listen to the soothing tone of the clerk's voice.

After a few audible yet pleasant sounds emanating from the insertion of the card key, you enter your room, taking immediate notice of the spectacular view through the window. The high thread count of the sheets is apparent to the touch. A mint was left on the pillow. A little sign guaranteeing freshness sits next to a handwritten thank-you note from the housekeeper. You can't help but run your hands over the soft, plush towels. And, of course, the ends of the toilet paper are nicely folded into a point providing reassurance that the bathroom has been "sanitized for your protection."

Now consider this.

You get out of a cab at an old, roadside hotel. You hear the sounds of traffic and nearby construction. After paying for your room through the protective glass separating you from the clerk, you grab your key attached to a large plastic identifier. After dragging your bag up two flights of stairs, you make your way down an open corridor exposed to the elements.

You enter the room. It has a musty smell. There's a large "tube-style" television bolted to the cabinet on which it sits. The carpet looks like it was originally in a now-razed Vegas casino from the 1960s, and the bed permanently sags inward from overuse. The wallpaper sports a mixed display of fruit and flowers, and the bathroom smells of bleach. As you lay in the sunken center of the bed, you can hear the steady drips from the bathroom faucet in between the voices of people arguing in the room next door.

Location matters.

In the 1990s, Motel 6 began displaying solid black posters in their lobbies with the following phrase: "All hotel rooms look the same with the lights off." Although technically true, we do our most productive work in the light, and the surrounding environment is critical to its success. Regardless of what we do and where we are, it's almost impossible not to have a psychological and emotional experience based on the elements within that space.

We all have a tendency to spend much of our time in some very unproductive locations loaded with distractions. Both my home and office are decorated with purposeful, tangible aesthetics intended to improve my mood and make me "feel" more creative and motivated. However, people, the fridge, the television, and sometimes the dog, frequently interrupt my stream of thought and thus, my productivity.

To effectively complete tasks with higher levels of both creativity and imagination, I try to do it at one of my "sweet spots," a secondary place where I can disconnect from the world and feel completely relaxed and energized. For me, these places tend to revolve around local restaurants and vacation spots. Restaurants work well when I'm simply trying to focus on something specific. People generally ignore me, and the surrounding activity serves as white noise to help me stay mentally locked in on the task at hand. I wrote my entire doctoral dissertation at Applebee's, my last couple of books at Chili's, and much of this book at Subway. Most of the creative thinking, outlining, and research were done while relaxing poolside in Las Vegas. For whatever reason, these places work for me.

Our emotions directly affect our focus and creativity. A secret to productive thinking is the ability to identify those personal, five-star sweet spots where you can feel your emotional, intuitive best. Go there—whenever you can—when you want to be most focused, energized, and creative.

6. Slow Down and Think!

Quick question: If you go somewhere and realize that you've forgotten or lost your smartphone, do you feel a little empty, naked, or lost?

Instead of actually being present in the moment and enjoying it, being without our smartphones can make us feel like a bunch of drug addicts in withdrawal. We also require our smartphones to keep doing more and more. In the last few years alone there have been thousands of "productivity" apps created for the smartphone to help us squeeze even more out of it. I find myself missing the "days of old" when a phone was actually just used to make a call…and that's all.

Life happens fast enough without the need to purposefully cram more into every second and find ways to accelerate it. Remember, our mental energy is finite. Our brains are easily filled and our energy depleted. Combine this with the fact that time is constant and doesn't change (there are always 60 seconds in a minute, 60 minutes in an hour, and 24 hours in a day), it doesn't take much before you start feeling overwhelmed.

This is when a little focus math might help ensure you're doing the right things with your limited time and not trying to do everything, which is a recipe for failure. Consider the following formula that simply represents the total amount of time available in our lives—any "non-math" type people please remain calm as this will be easy:[11]

$$\mathbf{+|>} \quad = \quad \mathbf{<|-}$$

This formula basically illustrates that both mental energy and time are constants. If you wish to add something new to your "to do" pile (+) or are already doing something that you wish to do more of (>), then you must either reduce (<) or eliminate (−) something else in order to keep the equation balanced (a.k.a., something has to give for the non-math people). The time/activities formula always needs to remain in balance, otherwise stress, feelings of being overwhelmed, declines in the quality of your efforts, and a host of other

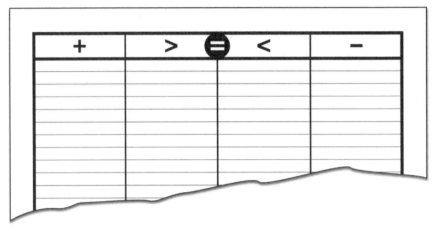

Figure 7-1. A sample worksheet.

non-desired outcomes will ensue.

The easiest way to use this formula to assess your daily activity (perhaps while you P.E.E.) is to utilize a simple worksheet where activity, tasks, etc. are laid out according to their value and placed in the appropriate column (see Figure 7-1).

To download a free PDF file of the above worksheet, visit my website at *www.adpaustian.com/resources*.

This worksheet will provide a visual assessment of how you are currently using two of your most valuable commodities: time and mental energy. Modify your life accordingly and take control of both.

These six simple actions have had a huge positive effect not only on my productivity and the quality of my work, but also on my ability to focus. They allow me to focus intently for scheduled blocks of time while putting my best efforts toward the desired task. Although they might be simple in concept, they aren't always simple to do. They require discipline and, most of all, frequent practice. Start by creating a routine and setting aside small amounts of time. Learn to appreciate the value and importance of time, and you might be shocked at what you can accomplish during only 30 minutes a day when highly focused.

One of the greatest purple people eaters in our lives is anything and everything we allow to distract us and ultimately keep us from being focused. According to a study at the University of California, Irvine, the average office worker is interrupted or distracted about every three minutes. However, the same study also showed it takes that same worker about 23 minutes to return to a task after the interruption.[12] There's an old Polish proverb: *"Not my monkeys, not my circus."* The proverb loosely means, "not my problem." It's easy to allow so many things outside of our control to dictate how we live our lives including things in other people's lives within which we have "chosen" to interject ourselves. We unnecessarily focus on them and try to be the ringmaster of it all. This only serves to eat up precious, limited time, increase stress, and most of all, destroys our ability to focus.

As we age, it will also frequently "feel" like time is accelerating. We feel this way because the perception of time is relative. For example, if you are 50 years old and going to Disney World in six months, that six-month span of time represents a mere 1% of your total lifespan which is why it feels like the trip isn't that far down the road and will be here in no time. However, if you are five years old that same six months represents 10% of your total lifespan which is why it feels like that magical day takes "forever" to get here. Also, as you've gotten older you've managed to add a whole bunch of stuff to the left side of your time/activity formula which tends to occupy and distract your mind, thus reducing the amount of mental energy that can be dedicated to thinking about Mickey Mouse...or the crowds, cost, and chaos that come with him.

However, time isn't relative; it's constant, and we can easily lose sight of how long one minute truly is until we focus on it. Grab that smartphone and set its timer for one minute. Hit start and close your eyes. That one minute will feel like an eternity. If you truly want that specific "someday" to happen, learn to focus and gain control of all of the minutes in your life.

Taking a Few More Steps:

• Download the time assessment worksheet at *www.adpaustian.com/ resources*. Follow the instructions and *honestly* complete it. In order to spend more time doing the things you want, where do you have to make changes? Repeat this exercise regularly (at least monthly).

• Do you find yourself frequently reacting to problems or situations? Life is full and complicated, but we all have to allow time to think and process. Regardless of how busy life can get, challenge yourself to set aside quiet time to think, process, and plan—in other words, allow yourself the time to prepare for life's storms before they hit.

• Keep a journal consisting of *one full week's worth* of decisions. Document any and all decisions you make from the most mundane (e.g., what clothes to wear, what food to eat, etc.) to the most critical and important (e.g., financially-related, strategic, etc.). Following the week, look back through the list and determine which decisions could become routine with little or no thought given to them. Predetermine how those decisions will be made ahead of time and shift your focus toward those most important. You should feel a greater sense of energy when addressing them.

• Think about what you want long-term…what you really want. Then, break that down into very tiny, incremental steps. Once done, while keeping in mind all of the aspects mentioned above, focus exclusively on achieving that first step, and only that first step. After you accomplish it, move on to the next. Not only does this keep your expectations at manageable levels, it keeps happiness within reach.

• Simplify your life. Simplicity provides greater certainty and decreases stress.

Noses, the Grindstone, and a Guy Named Bob

The Entitlement of Opportunity

As a child, I was taught to strive for the American Dream, something you could "earn" through hard work and persistence. It wasn't something you received because you were an American; the only aspect of the American Dream someone was actually entitled to was the *opportunity* to work hard for it. People who migrated to this country understood the hard work necessary for achievement and believed it was worth the sacrifice. The opportunity to work hard for something inspired the high levels of motivation needed to succeed.

I grew up in a middle-class home where my parents often said things like "whatever you do, put your all into it," or "if you borrow something from someone, always return it in better condition than you received it." Their actions reflected these views; they walked the talk and taught by example. These mini-lessons on work ethic throughout my childhood became the mortar that formed the foundation of who I am today. Keeping my "nose to the grindstone" became the mantra of my personal life.

As a teenager, I had many hobbies, interests, and lofty dreams, combined with a strong desire for success. Yet, my interests and plans required money I didn't have and needed to earn. So, I did whatever I had to do: mowing lawns, detasseling corn, picking tomatoes, babysitting, bagging groceries, painting, washing cars, selling nails (picked up from the ground at construction sites—with permission—after they were dropped), and even selling fireworks. My friend and I purchased fireworks in Missouri, where they were legal, and brought them back to Iowa, where they weren't so legal. We then sold them at a considerable profit. Although we ultimately did get in a little trouble with the police, one had to admire our teenage entrepreneurial spirit.

This ingrained work ethic also carried over into my schoolwork, which resulted in good grades made possible because I consistently met and/or exceeded the amount of assigned work.

Today, when I think about work ethic, I have to ask what it means now, how it fits into people's lives, and how it impacts the process of imagineering. Creating good ideas through effective imagineering is hard work and requires a determined work ethic. If it was easy, everyone would be doing it, and all ideas would be good (which then waters down the meaning of good so ultimately nothing is good). Work ethic has been defined as "a set of values based on the moral virtues of hard work and diligence."[1] However, the term "moral virtues" seems to imply a sense of universal understanding as to the meanings of both hard work and diligence. People can be very busy, working very hard for a long period of time, yet really do nothing of value to improve their lives or help them obtain the "American Dream" as they perceive it. However, according to this definition, they might be viewed as having a strong work ethic despite the outcomes. But, do they really?

A Slightly Different View

I have an alternate definition, one I believe captures the essence of work ethic. To me, work ethic can be defined as simply "doing whatever you need to do in order to get whatever you want." When people persistently and consistently take all of the steps necessary to reach a desired level of personal success, they will exhibit the degree of work ethic required to get there.

For example, if someone is content making $20,000 per year by steadily working a full-time job that pays $10 per hour, then that person is exhibiting the required level of work ethic to achieve this desired level of success. Although others may not agree that this person exhibits a strong work ethic, success is a very relative term, and its definition often depends upon the views of the person evaluating it. Many variables go into one's personal view of success that might run contrary to the view of society at large. The key is to realize that as long as you do whatever you need to achieve your desired level of success, you are exhibiting the necessary degree of work ethic regardless of what others may think.

However, when people desire something more, something they currently don't have, or their expectations of success increase, they must work harder to achieve this higher level of "want" and exhibit a stronger work ethic. Going to the Moon required an incredible level of work and commitment on the part of all 400,000 involved, as their lives were totally upended during the 1960s. According to Fred Haise, Lunar Module Pilot for Apollo 13,

> *"It was a hectic time, it was hard on people...up until, probably, literally the lunar landings, a little beyond—it was seven days a week, 24 hours a day...one year between '67 and '68 there was only one day off all year...Christmas."*

Walt Cunningham, Lunar Module Pilot for Apollo 7, put Apollo's success squarely on the 400,000 NASA personnel and contractors.

> *"[E]very single engineer owned that program. When I say 'owned' it, it was as important to him as it was to me and what I was doing. To this day, you can talk to them and they will let you know that it was their program and it wasn't going to fail because of something they did or didn't do. That was the kind of commitment and dedication that we had... The public today has kind of gotten away from that sense of biting off the impossible and going in and solving it regardless of how difficult it is..."* [2]

Alfred Worden, Command Module Pilot for Apollo 15, believed the success of the Apollo Program was totally the result of both work ethic and teamwork.

> *"I was assigned to work with the manufacturer to rework the engineering of the command module after the Apollo 1 fire. We had spent countless hours and days on the assembly line checking out the spacecraft during final assembly to ensure it was reliable for flight. Since these people were in the best position to know what, if anything, needed to be corrected or improved, I believed it was critical to get to know them as people, not just as workers. They were deeply committed to this project. As the guy who was going to fly this machine to the Moon, they knew I only wanted their very best effort. They gave it, and then some, while working tirelessly to correct many problems that could have had serious consequences during flight. The close partnership created by working together side-by-side allowed me to witness firsthand their passion, pride, and incredible work ethic."* [3]

In his book, *The 21 Irrefutable Laws of Leadership*, John Maxwell discusses the "Law of Process" whereby "Champions don't become champions in the ring—they are merely recognized there."[4] In other words, when we watch Michael Phelps win his 23rd Olympic gold medal swimming in Rio or Serena Williams win her 7th Wimbledon title, all we see are the results, not the countless hours spent getting there through practice and training. These champions made a decision many years earlier that they wanted to become champions, and then did what they needed to do to achieve that level of success.

The Story of Kurt

After a nice stint as an all-state high school quarterback in Cedar Rapids, Iowa, Kurt Warner was disappointed to learn that no Division I-A (FBS) college was willing to offer him a scholarship to play foot-ball. Shifting his focus, he accepted a partial scholarship to a smaller Division I-AA (FCS) school located about an hour's drive from home. Believing he would get to play early and often, he went to the University of Northern Iowa (UNI), where he would be relatively close to home and his family and friends could watch him play. He was wrong. After three long seasons riding the bench as a back-up quarterback, Warner finally won the starting position during his senior year. After a rocky start to the season, he ultimately earned Gateway Conference Offensive Player of the Year honors.

Warner believed his level of play that year would be good enough to earn a spot in the NFL draft, but once again, he was wrong and went undrafted. However, in 1994 he was given a chance to earn a spot as a free agent on the roster of the Green Bay Packers, but he was competing against a young Brett Favre, veteran Mark Brunell, and former Heisman Trophy winner Ty Detmer. He was cut from the team before the start of the season.

Shortly thereafter, Warner returned to Cedar Falls (where UNI is located) to work in a grocery store for $5.50 per hour stocking shelves and served as a graduate assistant coach for his former UNI football team. With no NFL team willing to give him a tryout, Warner signed with the Iowa Barnstormers in 1995, part of the Arena Football League. He experienced great success in arena football and was named to the AFL's First-Team All-Arena in both 1996 and 1997 after leading the Barnstormers to the Arena Bowl (the league championship game) both seasons.

Because of his hard work and success with the Barnstormers, he signed in 1998 with the NFL St. Louis Rams and then was sent to play for the NFL Europe Amsterdam Admirals, where he led the league in both touchdowns and passing yards. Following his season in Europe, he returned to the Rams and served as their third-string quarterback after beating out Will Furrer for the position.

At the beginning of the 1999 season, Warner became the second-string quarterback behind newly signed Trent Green. However, when Green tore his ACL in a preseason game, Warner finally had his opportunity to take over the starting role. With the support of his coach, Dick Vermeil, and fellow teammates, Warner had one of the best seasons in history for an NFL quarterback, throwing for over 4,300 yards, with 41 touchdown passes and a pass completion rate over 65%. The Rams went on to win Super Bowl XXXIV, and Warner was named 1999 NFL Most Valuable Player as well as Super Bowl MVP—one of only seven players, including such NFL greats as Bart Starr, Joe Namath, Terry Bradshaw, and Joe Montana, to receive both honors in the same year.[5,6]

Kurt Warner's journey to becoming a successful NFL quarterback demanded hard work and perseverance. He always knew what he wanted, but getting there wasn't easy. Unlike most NFL quarterbacks, who usually go through a simpler, more traditional process of career progression, Warner's path was significantly longer and much more difficult; he had to humble himself and repeatedly accept "lesser" roles in football just to stay connected to the sport.

Despite all of his personal setbacks (which also included financial and family problems), he maintained a strong belief in his potential and did what he needed to do to achieve his goal of playing in the NFL. In his book, *All Things Possible*, Warner said, "I believe that the Lord has a plan for each of us that's better than anything we can imagine—even if that plan isn't obvious to us at every stage." He went on to describe how God had prepared him for this process over a long period of time—in lower-profile locker rooms, the grocery store, playing in Europe, through all of his personal tragedies, and in spite of the people who doubted him along the way. "I realize now that I would not have been prepared for my big chance had it happened before it did."[7]

Some might say Kurt Warner was just lucky—he was in the right place at the right time. I think Kurt *was* lucky. To me, luck is nothing more than a fortuitous unplanned opportunity. We are all surrounded by luck, but whether or not we are "lucky" depends on whether we're able to take advantage of that luck when it presents itself. Through Kurt's hard work and persistence, he put himself in a position to take advantage of the fortuitous, unplanned opportunity of Trent Green's injury. Had Kurt not worked his way into the second string position, he couldn't have seized the starting role when Green was injured. Like Kurt, through hard work we can put ourselves in a position where we're ready to greet Lady Luck when she comes along.

My Experience

Over the years, I've had the opportunity to teach a large number of college students in a variety of business-related topics. Most recently, my efforts have been with fourth-year students in a capstone strategy class designed to tie all of their prior business courses together with an applied focus.

One of the topics that I stress is the necessity of being willing to do whatever is needed in the short-term to achieve what is wanted over the long-term. Today's world has become extremely competitive; unfortunately, a college degree doesn't necessarily guarantee a job to the person who thinks he deserves it. As a result, a student may have to do something he really doesn't "want" to do in the early stages of his career to position himself for that which he truly does want (like Kurt).

Most of my students seem to feel this principle doesn't apply to them, and they will have to figure that out on their own. However, every now and then one of them will listen and take this advice to heart. In an email, Melissa, one of my former students, shared her job-seeking experience after she graduated:

> *"Watching my friends graduate at a time when it's so hard to get a job, it saddens me to see the reactive state with which so many people have become comfortable. When they don't find their dream job within a couple of days searching, they attribute their difficulty to the economy and the horrors of the job market. It stuns me the lack of effort they put forth. I graduated with excellent grades and a degree in information systems. After graduating in May, I applied for jobs daily for months until I received my first interview at the end of August. I nailed the interview and landed my sweet gig...in reception.*

Never in a million years did I anticipate that all my hard work in school with a "legitimate" major would get me is a job that I could have just as easily received with a high school diploma. But I wanted a job, so I took it. With slightly bruised pride, I worked very hard to prove both my potential and desire to do bigger and better things. Ten months later, I was promoted into an entry-level job in my field of study...I was proactive, and finally felt as though I had direction in life.

The most astonishing thing to me is how all of my friends still complain almost a year later about being unemployed after graduation. Yet these are the very same people who turned down the opportunity to submit their résumés for my old reception position after my promotion because it was somehow beneath them." [8]

Declining Drive

As nearly any educator can attest, over the last couple of decades, students have changed. As a professor, beginning with my very first class, I have always given students the opportunity to write an extra-credit research paper should they decide to do so. The paper requirements and constraints were detailed in the course syllabus from the first day of class, and the paper was due the last day of class, giving students approximately four months to prepare it. The paper was worth up to an additional 10% of the final point total, so their final course grade could potentially improve by a full letter grade.

When I first began doing this, I received about 28 papers out of a class of 30 students, some from students who were already earning an "A" for the class. Most recently, in a class of 30 students, the number of papers received had declined to only 2, or about 7% of what I received 20 years ago (and these were turned in by students already earning an "A"). This change represents a steady decline of more than 90% over two generations of students—students who are now apparently unwilling to do "whatever they need to do" in order to guarantee a higher final course grade.

Over a span of five years, I also performed a non-scientific study with my fourth-year students in the strategy classes discussed earlier. During discussions of creative and innovative thinking, I had students break into groups of two and gave each group a bag containing a number of miscellaneous items, including some Lego blocks, a penny, a nut, a bolt and a washer, a marble, wire nut, and a wooden spool. Each group was to divide all of the items into two distinct groups (i.e., Lego blocks, not Lego blocks—see Figure 8-1 below). They were to do this as many times as they could within 20 minutes. I also passed out a solution sheet for each two-person group that restated the instructions for the exercise—in boldface—and included a number of lines so they could record their solutions. What I didn't tell them was that the number of solution lines per sheet gradually increased from 20 to 80 as the sheets were distributed across the room.

Figure 8-1. Parts in exercise separated into Legos–Not Legos.

An interesting phenomenon occurred. Students completed the exercise and came up with as many solutions as possible up to the number of lines they were provided. In only one single instance (out of more than 300 hundred total groups) did a group go beyond the number of provided lines. Although there were many fully completed sheets ranging from 20 to 80 solutions, once a group filled the respective number of lines, imagineering stopped and group discussion shifted to random topics while they played around with the items.

Although the instructions were clear and it was stressed each group needed to come up with as *many* solutions as possible, nearly all groups only worked up to the self-imposed, perceived limits of the exercise. Some groups arrived at 20 solutions, others 40, and still others 80, yet for the most part, students seemed equally content with the amount of effort they invested in the exercise (ironically, the list of solutions I have compiled over the years includes over 500 possibilities).

These classroom examples provide anecdotal evidence as to how some people approach work today. I have also observed similar behavior patterns with many workers at all levels (e.g., unwilling to stay longer if the job isn't finished, only doing that which they are asked to do, and/or failing to take initiative outside of assigned tasks). These experiences, combined with similar stories shared by colleagues, lead me to believe there is a cultural pattern of a declining work ethic in comparison to what people say they want and desire.

It Keeps Going and Going

Imagineering isn't easy, and it often requires hard work to develop a good idea. Once an idea is realized, it frequently takes even more work to ensure it's a good one.

Therefore, the concept of work ethic also implies a certain level of persistence. Sometimes getting what you want may require extra time and effort doing what you need to do in order to get there. Persistence also requires steering clear of purple people eaters such as vacuums. History is replete with examples of successful people who displayed extraordinarily high levels of perseverance and persistence to achieve their goals and dreams.

For example, James Dyson spent 15 years and developed over 5,100 prototypes to finally complete the design for his revolutionary vacuum cleaner (one that truly sucks compared to others in the market—pardon the pun). When no other manufacturer would take it on as part of an existing product line, Dyson launched Dyson Limited to manufacture and distribute it. Dyson's idea is now one of the top selling vacuums in the United States.[9]

Chester Carlson's idea for electrophotography was shot down more than twenty times by companies such as IBM and Kodak. Through his perseverance, however, he was able to finally enter into an agreement with a small photo-paper company called Haloid (later known as Xerox). Twenty-one years after initially inventing electrophotography (now known as xerography), the first convenient office copier using the technology was unveiled.[10] Imagine an office today without a photocopier or laser printer!

Dr. Seuss (a.k.a. Theodor Seuss Geisel) peddled his first children's book, *Mulberry Street*, to 27 different publishers only to be rejected 27 times. It wasn't until he bumped into an old college friend who happened to work at Vanguard Press, a division of Houghton Mifflin, that he was able to get his illustrations and manuscript in front of some key decision makers. Vanguard ultimately published the book, which was well-received, and jump-started a career that resulted in 44 children's books, nearly 30 of which have been adapted for television or video.[11]

Books, movies, and television shows are loaded with examples of people like Dyson, Carlson, and Geisel (Dr. Seuss). Why do we like these stories so much? Could it be that the people in these stories exhibit such a high degree of perseverance that one can only dream of reaching that level? Or, could it be that most people believe this kind of work ethic and persistence only exists in books or movies so the entire concept is fiction?

What If?

I remember the day my son purchased his first car, an old Chevy Cavalier with a ton of miles and a salvage title. Shortly after the purchase, he was in my driveway washing and waxing it. He even scrubbed the interior, washed the carpet, and added a nice perfume scent. When I mentioned that I didn't ever recall him washing and waxing my car during all of the time he had driven it, he replied, "That car wasn't mine."

In reality, I believe most people share this same perspective. There's an old saying: "No one ever waxes a rental car." When there is little or no personal investment, it's really hard to place a lot of value on something. Sometimes we can see it in children and how they take care of their "free" stuff while growing up, college students who haven't made a personal financial investment into their education (or can't feel the pain of their student loans while still in school), and employees who are allowed to simply "exist" within a company and not pull their weight.

Unlike years past, we have a growing number of people who receive many of their "wants" without making a personal investment into acquiring them. This process of entitlement or guaranteed benefit leaves out the most important aspect of acquisition: the hard work associated with getting there. The process of creating personal value and self-esteem is why a strong work ethic is so important for the long-term growth of people. The feelings of accomplishment, a goal achieved, and personal success are what have driven this country and our entrepreneurial spirit since the beginning.

Leaders today at all levels—from parents to teachers to office managers to production supervisors to the President of the United States—should facilitate an environment of self-growth, support the efforts of other people, and provide help and inspirational guidance along the way. The more success people have "doing whatever they need to do to get whatever they want," the more they will be driven to repeat the process while also helping others get what they want. Their sense of self-esteem and value will drive their personal growth and encourage them to reach higher levels of success. Their passion for work will become contagious and spread to others around them.

Someone once said, "There are no menial jobs, only menial attitudes." In his book *Brain Droppings*, comedian George Carlin observes, "Most people work just hard enough not to get fired and get paid just enough money not to quit."[12] What if people suddenly realized there is honor in all work? What if they saw the value of hard work as a way to give? What if everyone in your family, company, or organization had a stronger work ethic? What if they all worked a little bit harder, and their noses moved a little closer to the grindstone? What if?

Big Leaps and Baby Steps – The Role of Leaders

I was once asked to teach a college capstone marketing class for two semesters for a professor on sabbatical. The students were fourth-year marketing majors, and the purpose of the class was to bring their marketing education together into a holistic, strategic context.

To provide them with "real world" marketing experience, I arranged to collaborate with a local business that manufactured and sold a high-end product line. The owner of the business agreed to let teams of five students develop strategic marketing plans and create promotional campaigns for his business (which was especially appropriate since college-aged students are one of the primary target audiences of his product line).

The culmination of the project was a formal presentation made by each team before a panel of judges, which consisted of the company owner, two experienced local marketing professionals, and me. The owner promised to use all or part of the winning team's approach to market the company's product and agreed to give each student of the winning team a $250 Visa gift card. Students had both financial and résumé-boosting incentives to perform well. The owner also agreed to cover any out-of-pocket presentation expenses up to $100 per team.

As the judges listened to the presentations that first semester, it quickly became apparent the students lacked passion and motivation for the project. Members of the panel frequently commented about how difficult it was to select a winning approach. In general, the projects lacked creativity; some ideas seemed to be little more than knock-offs of existing approaches (or tired, old ones). It was clear the majority of students hadn't invested enough time on the projects or the presentations, some of which bordered on unprofessional.

After some convincing and a little pleading on my part, the owner agreed to sponsor the project again the second semester. This time, I was determined not to make any assumptions regarding the skills of college students who were about to graduate and to provide them with the tools necessary to be successful.

I devoted several class sessions to the concepts of strategic creativity, professional presentation, influence, and risk-taking. Instead of merely assuming that students were frequently meeting outside of class to work on their projects (as the little time I had provided in class that first semester didn't seem to be well utilized), I set aside a greater amount of class time for them to work on their projects. I actively engaged the groups and served as a resource.

The end result: an outcome worse than the first. Not only were the presentations even more disappointing, the panel members became concerned about the overall quality of potential marketing graduates of the institution. My excitement and anticipation about providing students with a great professional opportunity quickly shifted to embarrassment. Even worse, my views of today's college graduates were negatively altered in a significant way.

So What's the Problem?

Initially, I believed the problem rested squarely with the students: they lacked the motivation, work ethic, and creativity required to be successful. What student wouldn't want to earn some cash and gain additional résumé experience right before venturing into the world of job hunting? But, after some soul-searching, I realized that some of the problem resided with me. Perhaps I had made too many assumptions; perhaps I didn't create an inspiring environment for learning to take place. Naively, I had simply assumed what appealed to me would also appeal to them. In short, I failed to be an effective leader in the classroom.

Please don't misunderstand. I still believe the students didn't exhibit the necessary motivation, work ethic, and creativity to be successful. However, I have also come to realize—and better under-stand—my responsibility as the leader of the class was to create an environment that allows those traits to emerge. Unfortunately for all involved, it took two semesters and a bunch of wasted time and money for me to reach that level of understanding.

Motivation and the Influence of Environment

Research in the field of psychology indicates that human behavior (or action) is a function of three things: motivation, ability, and opportunity.[13] Motivation is an internal drive so people will typically have to want to do something in the first place. A leader can have the greatest success mobilizing desired behaviors by creating an environment that influences people's views related to the other two: personal ability and opportunity.

My students could have been motivated in a variety of ways, of course, ranging from a genuine interest to learn to simply enduring the process in order to obtain the necessary credits to graduate. At the time, I didn't ask (at least conceptually) whether they would rather enjoy or endure the class, nor did I create an environment around the exercise that was dynamic and engaging regardless of what motivated them, one where students believed they had the necessary prior knowledge or one where students could see the project's long-term value. Although the opportunity for learning was there, not all students recognized the opportunity. Combine this lack of recognition with the team aspect of the project, where varying levels of perception and motivation must come together, and the result is a recipe for an unproductive project environment and lackluster student performance.

Because of this experience, I now evaluate behavioral outcomes within a broader environmental context, one which includes both psychological and spatial elements. For example, think about a typical elementary classroom with walls totally covered with student work, learning tools, visuals, and positive, colorful images. What would happen to this environment if these were all suddenly removed? What would it do to the energy in the room? How would it affect students' feelings and motivation? For whatever reason, most classrooms in higher educational institutions are just like this—empty and void of positive influences. The same can also be said of a great many production floors, offices, and workspaces. To maximize outcomes as a leader, I must continuously assess how inspiring the work environment is in relation to assigned tasks to be completed by doers at all levels.

This is what inspired me to create the Celebrate! Innovation Exhibition (and ciWeek discussed in Chapter 5). The ciExhibition is an interactive learning environment that focuses on telling stories of creativity and imagination. The narratives are about people who were inspired to do things most other people believed impossible. Visitors learn how people from a variety of social, educational, economic, and ethnic groups have risen above the odds to take risks, motivate change, and add value to everyday life. The stories are told through the physical space where students and campus visitors are engulfed by tangible displays of personal innovation that serve as reminders of human potential. Whether it's an interactive exhibit of how the personal computer came to be, a visual timeline with artifacts that shows how communication evolved from the telegraph to the modern-day Internet, wall-to-wall visuals combined with video documentaries of the people who made a difference, or the biannual *Celebrate! Innovation Magazine*, the stories surround visitors with the intent of initiating conversations and prompting questions toward personal discovery.

The Big Three

Creative leadership, though, requires more than just an understanding of environment. Many other intangible aspects of leadership directly influence the behavior and output of others. A quick perusal of the leadership section at Barnes & Noble or a search on Amazon (which yielded over a quarter-million resources on leadership the last time I looked) can reveal a nearly unlimited supply of advice. However, my experience suggests three core leadership principles which, when applied regularly over time, have helped me connect with people and impact their level of motivation: "Get Behind and Push," "Walk the Talk," and "Think Big, Execute Small."

Figure 8-2. Students enjoying an exhibit from the History of the Personal Computer portion of the Celebrate! Innovation Exhibition.

Get Behind and Push

Successful leaders must first be successful followers. Only leaders who know how to take direction, acknowledge responsibility, and accept accountability for their actions can truly understand leadership and its impact.

I learned this very early in life. When I was a child, it was still socially acceptable to receive a spanking whenever I exhibited some unacceptable behavior. Fortunately, in my case, the spankings administered by my parents hurt just enough to remind me that my behavior was not acceptable and would not be tolerated. My parents weren't at all physically or mentally abusive; however, their actions played a large part in teaching me how to follow directions and gain the understanding that I would be held responsible and accountable for my actions.

Along with being a successful follower, a real leader also knows it's the doers that get the job done. So, point the direction (preferably one that's a big leap) and get out of the way while supporting the process. Strong leaders push others toward success. They recognize that when their people are successful, they too are successful, and success is mutual. Real leaders activate the law of reciprocity: when you help enough people achieve their goals, your own goals cannot help but be met. Put simply, givers get. Succeeding faster only requires one thing: a selfless devotion to helping other people achieve success by assisting them with contacts, resources, information, and guidance.

This system of support also includes understanding and connecting to others by caring and showing interest, by acknowledging and sharing their ups and downs, by treating them fairly and as equally important, and by viewing them as partners instead of subordinates. Strong leaders create an almost "family-like" supportive environment where relationships are necessary, trust is built, and it's in everyone's long-term interest to help others succeed. They never put themselves in a position to take something of value from those they are leading, which would ultimately devalue effort and kill motivation.

Walk the Talk

Leadership is a daily process, not a destination. Before you can effectively lead others, you must first lead yourself. In other words, a strong leader leads by example and knows their personal character will set the tone for everyone else. You must consistently display the character traits required by everyone to ensure success. Dependability, patience, self-discipline, integrity, confidence, and a strong work ethic become daily expectations of you. Others cannot be expected to do what you are unwilling to do, and a good leader knows a consistent, high level of character is critical, whether one "feels" like it every day or not. Character can't be faked. One's character is reflected when no one is watching, and others will see through insincerity.

At 211 degrees, water is just hot. At 212 degrees, it boils. One degree can make a huge difference. Not only should effective leaders set the bar of expectation, they should try to do "a little bit more" and consistently exceed expectations each and every time. Most people tend to value how others make them feel and will attempt to acquire the feelings they desire by associating themselves with those who exhibit them. We like to be around others who make us feel better about ourselves. By accepting a leadership role, you commit to a higher standard, one that not only requires a strong character but also demands a positive attitude.

A story of two bricklayers illustrates this concept. One day, a pedestrian stopped to admire the skill of two men who were laying bricks. She asked the first bricklayer, "What are you making?" In a somewhat gruff voice, the bricklayer responded, "About $20 an hour." At a loss for words, the pedestrian stepped over to the next bricklayer and asked, "Say, what are you making?" The second bricklayer happily replied, "I'm making the greatest cathedral in the world!"[14] Same activity, same question, two totally different responses. A positive attitude will change one's total perspective of something. A good leader chooses to see problems as opportunities to do great things versus mere labor.

If you have ever ridden a rollercoaster, you know a wide variety of attitudes are exhibited on any given ride. Some close their eyes, hold on for dear life, and can't wait for the ride to be over, while others ride with eyes wide open, arms outstretched, and love every second. Same ride, two entirely different emotional responses, but those in the latter group typically take the lead by sitting up front.

Attitude is a game changer. It often reflects the tone of leadership and dictates the response to failure. Babe Ruth had to strike out 1,330 times in order to hit 714 home runs (both once records in professional baseball) and lead the Yankees to multiple championships;[15] Walt Disney was fired from his newspaper job for a lack of creative ideas;[16] Thomas Edison was pulled out of school as a child after his schoolmaster called him "addle-minded" and "slow;"[17] Michael Jordan missed over 9,000 shots in his career, lost 300 games, and missed 26 final game-winning shots on his way to leading the Bulls to six NBA championships;[18] and Lee Iacocca, having been fired from Ford after 32 years of service, went on to lead Chrysler back to success after the company was on the brink of bankruptcy.[19]

Attitude is an outward expression of the heart. If you truly want others to be successful, maintaining a consistent positive attitude is paramount. People can easily become discouraged by any one of a large number of aspects in their lives. A positive attitude by those in charge—as well as a positive environment—can help them overcome those feelings and develop a renewed sense of energy. Strong leaders strive to exhibit a positive attitude every day to help others exhibit one on most days.

In the spirit of one of my favorite books, John Maxwell's *21 Irrefutable Laws of Leadership*, I have developed my own personal leadership laws with the sole intent of creating and maintaining a positive attitude and work environment. Many of these began as simple ideas and were shaped over the years through a lot of trial and error:

• **Always think "Big Picture."** I will develop genuine holistic problem-solving skills, think creatively, and visualize all aspects of solutions before implementation.

• **Be passionate about what I am going to do, and then do it.** To build trust in others (which is necessary to accomplish anything), I will always exhibit a high degree of passion, energy, honesty, integrity, and follow-through.

• **Never expect more than I am willing to do myself.** I cannot expect people to go the extra mile, put in overtime, or do whatever it takes to be successful if I am not willing to do the same on a regular, consistent basis.

• **Always be fair and balanced.** I will strive to never show favoritism and always treat others equally regardless of status or position. Being fair and equal to everyone builds trust and loyalty and creates a sense of value.

• **Be approachable at all times.** People must feel like they can come to me when they need or want to do so. I will not set up any form of "gatekeeper" to control access to me, which would impede communication.

• **Manage through interaction.** I will make an effort to walk around and interact with others on a regular basis to keep me connected to the environment I am leading. I want others to see me as approachable and accessible.

• **Be consistent.** I will remain committed to staying even-tempered and displaying stability, thus modeling and promoting consistency in the decision-making process. People will always know what to expect, which will serve to minimize anxiety.

• **Surround myself with great people.** I will surround myself with and help develop those who possess and exhibit strengths greater than myself. A good leader acknowledges weaknesses and offsets them by partnering with others who are strong in those areas.

• **Share responsibility.** I will truly delegate responsibility, which serves to build value within people. I will not micromanage, yet I will continue to provide support when needed.

• **Give lots of credit—take very little.** As a leader, I am the ship's navigator, not its engine. I will never put myself in a position to take something of value from the doers I am leading, which would only serve to devalue their efforts and kill their motivation.

Think Big, Execute Small

Too often, people have a tendency to view the primary role of leadership as having and setting long-term direction, while letting others figure out how to get there. However, effective leaders are not only able to visualize which mountain to climb but also the individual steps necessary to climb it.

In the 1991 comedy *What About Bob?* Bill Murray plays Bob Wiley, a character suffering from some serious "issues" (the clinical diagnosis given in the movie was *an extreme case of multi-phobic personality characterized by acute separation anxiety*). When Bob's current psychiatrist pawns him off on Dr. Leo Marvin, an egotistical psychiatrist played by Richard Dreyfuss, Bob shows up at Dr. Marvin's office for an initial interview. As Dr. Marvin is getting ready to leave on vacation for a month, he shoves a copy of his new book, *Baby Steps*, into Bob's hands and sends him on his way. The premise of the book is to help people achieve larger goals by visualizing much smaller, reasonable goals and to then take a series of successive baby steps to get there. To the eventual dismay of Dr. Marvin, Bob totally takes the doctor's words to heart. He is able to visualize and take each necessary, yet very difficult, step toward "sharing" Dr. Marvin's vacation with his family. Bob's actions include walking to the bus terminal, getting on the bus, riding the bus to Lake Winnipesaukee in New Hampshire, finding Dr. Marvin by yelling for him in the middle of town, and then hijacking Dr. Marvin's book interview with *Good Morning America*. Bob humorously "baby steps" his way into every aspect of Dr. Marvin's life and eventual psychotic breakdown.[20]

Although they desire a different outcome, strong leaders are like Bob. They are able to "see" a big leap, some potential great outcome or challenging opportunity, and then visualize and implement each baby step necessary to achieve it. With laser-like focus, they accomplish each required step in sequence while keeping the big picture and ultimate outcome in mind the entire time. They realize that 20% of their effort accounts for 80% of their success (Pareto's

Principle) so they don't allow themselves to be overcome by vacuums, irrelevant daily minutia, and a variety of other purple people eaters. Able to manage many steps simultaneously while keeping the appropriate priority on each, leaders also recognize forward progress is a process. They are patient; sometimes great things may take considerable time to accomplish. In the Old Testament of the Bible, King Solomon says, "It is better to finish something than to start it. It is better to be patient than to be proud."[21]

Strong leaders will assemble great teams of doers who are able to execute. They will find and nurture those who can work both individually and collaboratively. They know that individual effort impacts the outcome of the entire group, so leaders are willing to work with doers to improve individual performance. Effective leaders are also willing to reorganize tasks and people to gain maximum output or remove people altogether if necessary.

President Kennedy saw a huge leap and knew it would take a massive commitment by the entire country to take a quarter-million baby steps to reach the Moon. A great team of focused leaders both in and out of NASA were chosen to help and support the over 400,000 doers necessary to take those steps. It was an enormous process where each strategic, successful step made within the larger scope of the program inspired everyone to take the next.

Imagine a snow globe. As long as each snowflake continues to fall, the desired effect is achieved. Sometimes, however, after the "snow" settles, the globe needs a good shake to reenergize it and keep things moving. Strong leaders are snow globe shakers. Have you shaken yours recently?

Taking a Few More Steps:

• Think about some of the various "environments" (workplace, home, etc.) where you spend the bulk of your time. How is imagineering either stifled or encouraged? If stifled, how can you improve the environment to do your best work?

• Think of an actual person or fictional character who lives in a very creative environment which appeals to you. Using this a reference, how could you modify your current environment to become similar?

• Which people in your life are the most supportive through encouragement and honesty? Spend more time with them.

• List five areas/tasks where you typically procrastinate. What is it about them you don't like doing?

• List five things you enjoy doing. What is it about them you enjoy? Create a plan that includes more of these things and less from the list above.

Remember the Alamoon!

"That's one small step for a man; one giant leap for mankind."

> – Neil A. Armstrong, Commander, Apollo 11
> and the first man to walk on the Moon

*"Whoopie! Man, that may have been a small one for Neil,
but it's a long one for me."*

> – Charles "Pete" Conrad, Jr., Commander, Apollo 12 and the third
> man to walk on the Moon (and the shortest Apollo astronaut).

July 20, 1969

I will never forget July 20, 1969. I was watching CBS News anchor
Walter Cronkite shed a tear as Neil Armstrong placed his boot on
the Moon for the first time. As a young child, I had no real concept
of the significance of the event. I just thought it was cool he was
on the Moon.

However "cool" it may have been to watch, that single event
inspired an entire generation—my generation. It showed us that we

could do almost anything if we worked hard enough and put our minds to it. It served as the driving force behind much of what we accomplished for the next few decades. Whenever a project or business activity reached a stumbling point or hurdle, the common phrase became "What, we can put a man on the Moon, but we can't...?"

It's also what personally motivated me to excel in school and focus on math, science, and electronics. In fact, as a high school kid in the late 1970s, I was developing personal computers and dabbling in robotics all because of the inspiration I received from the Apollo program and the astronauts who served as my personal rock stars. (Ok, so I was a bit of a nerd.)

Flash forward to March 5, 2010. We were hosting our first annual Celebrate! Innovation Week (ciWeek). The theme for the week was the Apollo Moon missions. The keynote speaker for the event was Alan Bean, Apollo 12 and Skylab III astronaut and the fourth man to walk on the moon.

I chose the Apollo program as the theme for our first Celebrate! Innovation Week because I believe it serves as one of the greatest examples of imagineering in the history of mankind. It was a massive project, to say the least. An incredible amount of technology had to be created for it, and by the first Moon landing, approximately 1 in 200 people in the country's workforce was in some way involved in the project (actually closer to 1 in 150 if the teenagers are pulled out of the total workforce numbers).[1]

Part of the imagineering was the development of the Saturn V—a rocket that towered 363 feet, weighed about 6.2 million pounds, and featured over 2 million functioning parts, all controlled by a computer with less power than what can be found in a nice toaster today. With the fuel alone, a car getting 20 mpg could drive 20 million miles, or 800 times around the Earth. At lift-off, the engines of the Saturn V generated 160 million horsepower (the

Figure 9-1. Alan Bean and I during ciWeek 1 in 2010.

equivalent of 85 Hoover Dams or over 500 jet fighters), enabling three astronauts to travel a quarter-million miles to the Moon, a heavenly body that is in constant motion in relationship to the Earth. The entire feat was accomplished by using the collective power of the human mind (even calculators didn't yet exist).[2]

Alan Bean's speech during ciWeek was inspiring as he detailed the events leading up to the successful completion of President Kennedy's vision. However, there was a repeated theme throughout his presentation about the lack of knowledge and understanding necessary to get someone to the Moon and how it required a large group of smart, but very average people working together to accomplish the mission. Since it hadn't been done before, no one knew exactly what to do, so they had to learn how to communicate effectively, solve problems as a team, manage large-scale projects, and most importantly maintain their drive and passion during failures and setbacks. To illustrate the "work-in-progress" and "learn-as-we-go" approach to the project, NASA stated in 1962 that the Apollo launch vehicle would be a "white-and-silver shaft jutting majestically 185 feet into the still-cool morning air." This statement underestimated the size of the eventual design by half.[3]

During our time together, I found Alan to be warm, friendly, intelligent, and talented, but he was also the type of person he talked about during his speeches, an average person (just like you and me) who had to figure things out on the fly while working with other people to accomplish something beyond what had only been previously imagined. Whether listening to his formal ciWeek presentations or just casually talking while eating a burger at lunch, Alan inspired me to reexamine my ideas about the roles of both leaders and followers.

Where Should the Focus Be?

I don't believe anyone today can dispute that we tend to put our leaders on pedestals and praise them for their greatness in overseeing a successful accomplishment. Whether it's Steve Jobs, Martin Luther King, Jr., Mother Teresa, Vince Lombardi, General Dwight D. Eisenhower, or George Washington, we have always admired leaders for the great outcomes occurring under their leadership. We buy the many books either written by them or about them, we fund studies to determine the common characteristics they share, and business colleges and others spend countless hours analyzing them. However, after listening to Alan Bean and thinking about the larger picture of the process, I think our focus has been misdirected.

I am not trying to downplay the importance of strong leadership. It takes a leader to set forth vision and direction, accept responsibility, and be accountable. It takes a leader to build a team capable of getting the job done. And, it takes a leader to serve as a unifying force during the entire process. Without President Kennedy's vision, we might never have gone to the Moon. Even after his death, he still served as a unifying force for those working on Apollo to remain steadfast during a difficult time in this country.

However, after thinking about the process as a whole, I believe we need to refocus our emphasis. We should be spending much more time focusing on the people who get the job done: the people who do everything necessary to actually achieve the vision set forth by the leader. These people, the "doers" for lack of a better term, are those who apply their education and skills to design and build the parts, the soldiers who put their lives on the line storming the beaches of foreign countries, the players who spend six days a week training hard for a three-hour game on Sunday, and the staffers who regularly work overtime to make sure reports are submitted by deadline. In the words of Alan Bean, these are the *"average people who come together to do something great."*

Without the doers, nothing would ever be accomplished. Therefore, the doers should be receiving far more of the credit and accolades for achieving the outcomes set forth by leaders. These doers made going to the Moon possible, they won the battles at Omaha Beach (D-Day) and Iwo Jima during WWII, they play and win the Super Bowl, and they get the presentation finished on time for the big meeting with the prospective client. The world needs doers…lots of them. Without them, nothing happens.

Doers Can Also Inspire

In 1836, Mexican General Santa Anna marched his army of more than 1,500 soldiers against the approximately 200 Texian soldiers defending the Alamo Mission. On their third advance, Santa Anna's army overwhelmed the Texians (later known as Texans after Texas joined the Union). They were either brutally slain or executed in the process and only the noncombatants were spared to share the story of the defeat with hopes of putting the fear of the Mexican Army in other Texians. However, the opposite occurred, and men flocked to join the Texian Army commanded by General Sam Houston. The rallying cry of "Remember the Alamo!" became the central motivator of the revived Texian Army, which went on to quickly defeat the Mexican Army and secure the independence of the Republic of Texas.

Although the doers at the Alamo (the soldiers and volunteers) were outnumbered seven to one and most certainly knew they would not survive repeated attacks by the Mexican Army, they demonstrated that with the right focus, a group of average people can do something extraordinary. Not only were they able to take out about a third of the Mexican Army during the battles, their sacrifice at the Alamo unified and inspired what was then a fragmented and disorganized Texian Army to go on and defeat Santa Anna.[4]

Like the more than 400,000 people who made Apollo possible, we need many people to accept the roles of doers. We need those who are task-intelligent and capable, united toward a common goal, willing to go above and beyond when needed, able to formulate and ask questions to understand and develop clarity of thought, open to necessary change, and willing to follow others. The process requires a coordinated effort on the part of doers who can effectively communicate and work as a team.

When Vince Lombardi was the coach of the Green Bay Packers, they relied heavily on a play known as the "Power Sweep." The play required players to pull out of their normal positions in order to block downfield, while the running back would run to wherever the other team wasn't. It wasn't a trick play or something very glitzy, but it was a very effective play that allowed the Packers to win five NFL championships. Lombardi once said, "It's my number one play because it requires all eleven men to play as one to make it succeed, and that's what *team* means."[5]

The Power of One

Regardless of the size of an organization, group, or team, success often comes down to the individual. It's been said that a single person can change the world, but a single person can also derail an entire process. A single player out of position can quickly cause even a simple play to fail.

While it's the role of leaders to focus on what the big puzzle will look like, it's the role of doers to focus on putting their specific pieces in place. If all doers are maximizing their personal effectiveness, the puzzle will successfully come together. For doers to be effective in their respective roles, they must understand the importance of proactive personal leadership and its direct impact on the five areas of personal health: physical, emotional, intellectual, spiritual, and financial.

Physical Health: *Directly Impacts Energy Level*

Numerous studies point to the relationship between physical health and productivity. Feeling good results in less down time, a better attitude, and more energy. There are many reasons why a growing number of companies are instituting wellness programs, with lower heath care costs being not the least of them. Physical health requires a life consisting of a controlled calorie diet, exercise, and the personal discipline to fight off the constant barrage of purple people eaters that try to distract all of us from achieving our goals.

Emotional Health: *Affects the Quality of Life and Work*

Numerous studies have also shown a direct relationship between physical health and emotional state (and vice versa). Stress, burnout, alcohol and drug abuse, depression, and a wide variety of other mental health concerns have a direct impact on productivity and serve as distractions and purple people eaters to effective outcomes. Engaging in positive activities and doing the things you enjoy will help you feel better about yourself and maintain a sense of balance in life.

Intellectual Health: *Continues Throughout Life*

Lifelong learning is a must, especially with life's rapid changes. Reading, taking classes, attending workshops and seminars, watching certain programming on television, and even playing games will not only improve your intellectual and problem-solving abilities, but can also be fun and relaxing, thereby improving both emotional and physical health. Lifelong learning also provides more ammunition from which to draw during the connection-making phase of imagineering.

Spiritual Health: *Creates Purpose and Virtue*

Believing in something greater gives purpose and direction to life. It creates the values, principles, and virtues that form the foundation for the decisions you make and direction you take throughout life. Realizing there's more to life than just you illuminates how your piece fits within life's big puzzle. It makes you a more effective member of your family, your team at work, and your community. For example, a thorough reading of our Declaration of Independence, Constitution, and Bill of Rights (all written by our founding fathers) displays a strong spiritual influence and belief in God.

Financial Health: *Allows the Freedom of Choice*

As evidenced by the relentless commercials for credit counselors, recent legal changes to "crack down on credit card company practices," and the rapidly growing burden of student loans, many people today face ever-increasing debt and the physical and emotional stress that comes with it. Those who are financially healthy know how to live within their means—which requires a budget—and work hard to both eliminate and stay out of debt. Financial health allows people the freedom to do the things they desire in life thus increasing personal joy and happiness.

Obviously, all of these areas are closely related and work hand-in-hand. Personal leadership is a holistic process that continually addresses all five areas of personal health simultaneously. The key to personal leadership is that it's proactive. You have to decide to be healthy. You have to take action toward having a better state of health. You have to follow through on that action. And, you have to evaluate the outcomes as you go and adjust as necessary. It's all about doing something to create active change and not just hoping for a different outcome.

Hope Isn't a Strategy

Of all the words and catch phrases popular today, the one I personally dislike the most is the misuse of the word "hope." Hope is not a strategy. It's not a vision or plan. It's not proactive or action-oriented. It's definitely not about effort or actually applying yourself. People "hope" for things to improve their lives. They "hope" for positive changes, and they "hope" that things will come their way or perhaps others will fix their problems.

In reality, I believe that hope—without a clear understanding of the larger effort needed—can serve as an excuse to do nothing at all, or at the very least, avoid the discipline and persistence required to accomplish anything of real substance. If we had "hoped" to get to the Moon, it would not have happened.

I recognize hope might be all that's possible for things outside our control. We hope a giant meteor doesn't crash into and destroy the Earth, and we hope cars won't cross over the centerline and hit us head on. But, hope alone is often inadequate for things within our control. In this regard, hope actually can become a paralyzing purple people eater if we simply wait on the actions of others or outside events to occur.

The people who have demonstrated great leadership over their lives and in the lives of others don't "hope," they "do." They are doers. They are the people who are willing to do their part and not simply wait on others to do theirs. They are willing to do the jobs (or tasks) that no one else really wants to do and do them with a positive attitude.

I had the honor of visiting the National World War II Museum in New Orleans. More than anything else, I took away the overwhelming sense of sacrifice on the part of everyone for the greater good of the country. Everyone—leaders, doers, ordinary citizens, those in combat and those who were not—helped fight Hitler's Nazis and the Japanese Empire. Everyone played an important role, whether they fought in battles, worked in factories, or rationed and conserved resources. People knew if they personally came up short, it could negatively affect the entire outcome. Personal leadership and discipline were the norm.

The same was true during the Apollo program. It's been reported many times that early in the program during a tour of one of NASA's facilities, John F. Kennedy asked a janitor what he did for NASA. He replied, "I'm putting a man on the Moon." Although the specifics of this story seem to vary, thus putting the factual nature of the account in question, my conversations with astronauts Alan

Bean, Gene Cernan, and Alfred Worden confirmed this very attitude exhibited the essence of NASA during the Apollo program. Alan said,

> *"Everyone knew their role and its importance towards accomplishing the goal. A group of average people were able to work together, each doing their part, to achieve something great."* [6]

When Gene Cernan, the last man to walk on the Moon during Apollo 17, spoke of going there, he always used "we" instead of "I," as if everyone who helped him get there went with him.[7] Al Worden, Command Module Pilot for Apollo 15, believed the relationship between the astronaut crews and the technicians who built and managed the spacecraft was absolutely critical. He said,

> *"I made it my business to know every tech on the production line, to the extent that I became the person they talked to when there was an issue. The people who built the spacecraft were so important and integral to the success of the program, I made it my personal policy to become part of **their** team."* [8]

Whether it's the legacy of the battle at the Alamo, the sacrifices made during World War II, or the motivated efforts of over 400,000 people in the Apollo program, history is loaded with positive examples of personal leadership and the necessary commitment demonstrated by all to accomplish shared goals. As we face growing challenges as a country, as companies, and as individuals, history needs to happen now. Hope isn't working, and personal leadership, responsibility, accountability, and effective imagineering should be placed front and center. People must recognize the importance and influence of their small piece of the larger puzzle and stop bemoaning the inadequacies of others. If every doer effectively fights off their personal purple people eaters and does their part, regardless of context, then every puzzle will come together. And perhaps to get there, we should all take on a new rallying cry, "Remember the Alamoon!"

Cows Aren't Always Cows: *An Image Perception Story*

Back in my early days of college, I worked as an assistant manager in a 24-hour grocery store. Being the lowest link on the managerial food chain, I was given the joyous responsibility to oversee the ever-popular 11 p.m. to 7 a.m. time slot (a shift which allowed me the opportunity to observe some incredibly interesting, sometimes disturbing people). During these hours, customer traffic would be slow, so we used the time to clean, stock and face merchandise, build product displays, and sometimes goof off.

One of my last responsibilities each morning was to check in the daily milk delivery. One day, for whatever reason, it dawned on me that ALL of our milk, regardless of brand, came off the same truck. I didn't yet fully understand the concepts of private labeling and multi-branding. At the time, though, I thought it was a brilliant and astute observation.

Our store carried three distinct "brands" of milk: Borden (the primary national brand with "Elsie the Cow" as its mascot), Parade (our store brand), and generic (boring white label with black text only). All of these brands came in a variety of flavors and types, including whole, 2%, 1%, and skim.

Since it seemed unusual to me that the same truck would deliver all of our milk, I felt compelled to ask the driver if he had to make several separate dairy stops in order to get it all. His response actually surprised me. "No," he said, "they all come from the same dairy; cows are cows."

He was right. There were no such things as Borden cows—although "Elsie" made a strong visual case otherwise—Parade cows, or "generic" cows. The delivery driver went on to explain the same exact milk is in each container; only the labels were different.

I then went on a mission (or a crusade if you will). I tried to "help" people by pointing out there was no difference between the three brands. My bosses didn't mind since we actually made higher margins on the cheaper milk. I distinctly remember a particular instance when I tried to explain this to an elderly lady who appeared to be on a fixed budget (remember what happens when you "ass-u-me"). When I saw her reach for the Borden milk, I quickly but politely suggested there was absolutely no difference between the brands and she could actually save money by purchasing the generic instead. I received a look I will never forget, one that suggested I was a combination of ignorant, stupid, and naïve, at best. She then proceeded to inform me that she had been drinking milk for about four times as long as I had been alive and I should basically mind my own business.

This was my first real lesson in branding and image: perceptions of reality are more important than reality itself, and it's perceptions that often drive people and their behavior. Despite having the exact same content, the packaging of the products created perceptions that were subsequently extended to the content itself. Elsie was a cartoon cow that had a very sweet and soft feminine voice in the Borden commercials. The consumer couldn't help but like Elsie, and her support of Borden milk added value to the product related to the perception of quality. It didn't matter that she was a cartoon. In the mind of the consumer she represented the ideals of being pure and wholesome. The Parade brand was only as good as the consumer's views of the store itself. As long as the consumer trusted the store, the consumer more than likely trusted its store-labeled products (although not to the same degree as beloved Elsie). As for the "generic" brand with its simple black text on a plain white background: despite no difference in product quality, why would anyone trust this product when given the choice? How good could it really be when no one was willing to put a name on it?

How Image Effects Doers

This same decision-making concept also applies to doers. In order to receive a desired degree of buy-in for a new idea or direction, a doer has to either believe in the idea itself, or at the very least the leader or seller of the idea. This buy-in is the result of perception related to the goodness of the idea (see Chapter 6) or the image of the person leading or selling it.

However, regardless of someone's current perception, it only takes one event, one issue, or one bad accusation to totally change how something is viewed. What if you found out, or someone simply made the accusation, that the cows that produced a particular brand's milk were being drugged, abused, or painfully electroshocked in order to produce higher quantities of milk? I doubt any cartoon cow in the world would be able to overcome that. What if it was reported that one of the store-labeled cans of beans had a dead mouse in it? Even if the store-labeled milk came from a completely different supplier than the beans, I'll bet that this would affect most people's view of all of the store-branded products. And as for the "generic," enough said.

Image is a lot like trust. You spend a lifetime building it and all it takes is one negative moment to tear down a lifetime's worth of effort. It's not only milk producers or even cola and shoe companies that have to be concerned about image. The Apollo 1 fire that killed three astronauts hurt the image of the program enough to almost kill it.

Whether you are a business professional, politician, athlete, parent, student, or someone trying to get by and make ends meet, your personal image is more important now than ever before. It can change in an instant, often before you even know it. I realize the concept of "personal image" may seem somewhat cheesy or unseemly to some people. Unlike past generations, where news traveled slowly and reputation was often formed by direct interactions with a few relatively local people, technology has transformed that. The sheer speed and ease by which information is disseminated has become a game changer.

Because people now tend to live at least part, if not most of their lives online through Facebook, LinkedIn, Twitter, Instagram, Flickr, YouTube, blogs, and a variety of other sites (including some, like my personal website, which may even seem "old-fashioned" to young audiences), a huge amount of online content has been created and shared, with or without permission. This content will tell people more about you than ever before: who you are, what you do, and why you matter. Whether you like it or not, this image is formed by the daily interactions you have with almost anyone, regardless of how formal the relationship. Although reputation is historically developed over time, it can now be shaped immediately, and search engines (like Google) have made it nearly impossible to "hide" from anything, good or bad.

In today's world, your personal image is the most important asset you own. It will directly affect both how people perceive you and your ideas in most aspects of life, whether their perception is shaped by your relationships with others, your job, or even your credit score. Leaders of all types and at all levels must have a positive personal image to effectively influence doers. If you take imagineering and personal leadership seriously, then you must do the same when it comes to your personal image.

So How is Image Created?

Others' perception of you is influenced by what they think about you, including how they look at you, how they feel when doing it, what they see, how they mentally categorize you in terms of others, and how (or if) they personally identify with you. In essence, your image—despite what you personally may believe it to be—is closely aligned to your reputation, which is interpreted by a variety of different people whose moods, perspectives, perceptions, attitudes, values, and feelings vary between them.

Although a great many brand experts claim you can proactively create your personal image and how people "see" you, I personally believe you are who you are, regardless of how you may present yourself to others. Although you may be able to fake being someone else for a little while, others will soon see through your charade, especially with the advent of rapidly changing technologies.

Your personal image today is now closely tied with your personal identity and all things you. No one can really hide from technology, so it's actually in your best interest to be transparent and authentic. In other words, be who you are and stop trying to be someone else. In a world driven by psychological shaping through the power of what "they" say ("they" being the product brands, celebrities, athletes, media, and anyone else who attempts to influence us today), the importance of being genuine has never been greater.

Also, through ever-improving search engines, it has become nearly impossible to separate our lives into distinct categories, such as personal and professional. Your life is now seen as one large clump of all things you, a holistic "digital personality" so to speak, and everyone has one (on the flipside, not showing up in web searches also says something about you). As a result, everything in your life now affects everything else, which can and will directly affect your future.

Since your digital personality represents all things online related to you, something as simple and innocent as a post to someone's blog, a post on Facebook, or an uploaded Instagram photo or YouTube video (or being "tagged" by someone else) can have permanent consequences. For example, as a college professor, I know of at least two specific cases in which straight "A" students were denied jobs because of some questionable aspects of their digital personalities. Unlike generations past, the choices we make today have a tendency to stick with us, possibly forever. At the rapidly growing rate of "online spread" (a term I use to describe the exponential spread of content and information through online reposting and tagging), even a few negative words about you— true or otherwise—can have disastrous effects.

This cultural shift in information transmission is not a "generational thing," only relegated to young people. Although much of the content on the Internet is still dominated by the Millennial and younger generations, older generations (Gen Xers and Boomers) have made huge gains. According to information found in the "Internet and American Life Project" of the PEW Research Center, the fastest growth in social network site traffic has come from older Internet users. Blogging has shown greater gains with the older generations, and older users are more likely to engage simultaneously in several different online activities, such as research, reading the news, visiting social sites, responding to email, and watching videos.[9]

Even an "old school" capitalist like Warren Buffet, CEO of Berkshire Hathaway, knows the importance of image and reputation.

Buffet was quoted in a *New York Times* article in regards to the dismissal of someone once seen as his potential successor. He said, "We can afford to lose money—even a lot of money. But we can't afford to lose reputation—even a shred of reputation."[10] If Warren Buffet is willing to lose substantial money rather than a shred of reputation, what does that say about the importance of personal image?

That is why now, more than ever before, part of personal leadership is doing what you can to take control of your digital personality and reputation. If you don't at least attempt to take control of it, others will do it for you, and you may not like the result.

Take the 2009 media frenzy that surrounded Tiger Woods, one of the first billionaire athletes many believed to have the perfect personal image since he turned professional in 1996. That image, crafted through his competitive golf game, the media, rapid online spread, endorsements, and positive word-of-mouth, came crashing down in an instant, and he and his game have never been the same since. His image was more than golf. It was also based heavily on his character, and his behavior away from the game of golf had directly impacted that part of his image.[11]

Toyota, a company founded in 1950, slowly built its market-leading image in automobile manufacturing based on strong perceptions of quality, safety, and excellent design. In the wake of major quality defects linked to a number of fatalities, these perceptions were seriously questioned in 2010 during the largest recall in the history of automobile manufacturing. Combined with production problems created by the 2011 Japanese tsunami and resulting nuclear reactor meltdown in Japan, Toyota still struggles to return to the same image perception it once had.[12,13] According to the J.D. Power and Associates Initial Quality Study, Toyota's brand quality perception dropped from sixth overall (top amongst mass-market brands) to 21st out of 33 brands.[14] The online spread of media stories and discussion boards related to the company's issues only served to accelerate this change in perception.

Building image and trust is a fragile process done DAILY, not in a day. Every act, experience, visual, and situation, however minor, impacts people's perception of how one's image is defined. Your image lives in the hearts and minds of everyone who comes in contact with you either directly or indirectly. Just as "Elsie" adds value to the Borden image by creating a personal connection with the consumer, the generic brand provides nothing to help create that connection, thereby adding to its long-term "risk" factor. Regardless of the context, people and doers want to be able to comfortably connect with you and trust the ideas you offer.

The longer an image is proactively built, the more likely it will be able to overcome periodic setbacks in perception, which can't be avoided (for example, a perception based on a false statement or accusation made by another). A daily, concerted effort to build an image focused on every detail and personal interaction will help most individuals or organizations overcome these setbacks.

How to Build Your Personal Image

As you develop and define a sense of who you are and ultimately your personal identity, there are a number of online (and some offline) things that can be done to directly influence the perception of others. It's never too late to start, and if you haven't, now would be a good time.

1. Decide how you want people to perceive you, see you, and generally feel about you. What characteristics make you unique? Where do you excel? Whatever you decide, it must be authentic and truly "you," or others will quickly see you as a fraud.

I can't think of a person in this world with a better personal image and reputation than my father. One would be hard-pressed to find anyone who has a negative thing to say about him. This image has been built over the span of his entire life, and he has served as a role model for me for all of mine. However, I am not my father. I can only be me, and any attempt to replicate his image and/or reputation would be seen as something I'm not. That's not to say that he hasn't influenced me; he has, and much of what I am today I owe directly to him. But, my personal image has to be mine. It has to represent who I truly am or wish to become, not someone I'm unable to be. Getting there is a process, and everything you do must be part of that.

2. Network. First and foremost, get out, meet, and connect with people. The best way to build a personal image is for people to get to know you personally. As long as you're genuine and attempt to put others first, the people you meet will tell others about you, and so goes the word-of-mouth process. As author Danny Beyer stated in his book, *The Ties That Bind: Networking with Style*:

> *"Networking is a basic human need. We need relationships, and networking is one way to build those relationships...The saying 'one person can change the world' is still true as long as that person has a powerful network to help...Do not be afraid to tell people what you want, and always be willing to help others."* [15]

2. Create an online home (such as a website or blog) that exhibits and provides evidence of your desired image. Remember that the quality of this site directly impacts the quality of your image (something that Starbucks, Pottery Barn, and Apple realized a long time ago). Provide relevant, value-added information, and most importantly, make sure it's accurate. Factual, grammatical, and spelling errors are all killers of a positive personal image. Your online home doesn't have to be business-related. It can be about your interests, hobbies, or activities. It just has to be about you.

3. Create profiles on social media sites (such as Facebook and LinkedIn) that support your desired image. These profiles and the information posted should be both personal and authentic, yet professional as well. Post only information, photos, and/or video that support your desired image, versus those you think are "cool" or "funny" and may actually cause long-term harm to your image. Think "Rated G," suitable for all audiences.

Also, since others can see your activity, be careful who you "friend" on these sites. You can become guilty by association by linking yourself to people with poor reputations or undesirable content. There are no prizes for having the most friends. Choose them wisely and regularly monitor their posts in order to be certain that those associated with you online are still supportive of your desired image.

4. Regularly communicate your image on the web through your social networking sites by posting comments to blogs or discussion forums, reviewing books or other relevant media, and asking for feedback, recommendations, testimonials, or comments related to your content (the latter builds credibility, but be careful what you ask for). It's always about quality and not quantity. Don't allow random posts to your sites and only enable comments to specific uploads or posts. Respond to negative comments and initiate a conversation regarding them, keeping the conversation positive and civil.

5. From a professional perspective, create original content that supports your online image through blogging, writing articles, adding supportive posts to others' work that is in line with your own desired image, creating video, and taking photographs. Through your life experiences you probably know more than you realize. In the offline world, write a book, volunteer your time, lecture or speak to classes and/or people at all levels, and join community-based and professional organizations. It also goes without saying that some "old-school" aspects of image building still apply, such as your style of dress and conduct. Personal behavior is still the biggest driver of personal image.

6. Always monitor! Stay on top of your social sites and what your "friends" are doing. What does Google say about you? Check regularly and make sure what's presented is accurate, and respond to false and negative portrayals in a positive, professional manner. Make sure the content or sites that you have some control over are near or at the top of a search of your name (use one or more of the many known search engine optimization techniques available). Set up Google Alerts to notify you when your name comes up on the web; over two-thirds of all web searches are conducted through Google.[16] Seek ongoing feedback from those you trust to ensure what you are communicating is perceived as intended.

Building an Organizational Image

From an organizational perspective, the creation of image requires all people, regardless of responsibility, to be properly trained and made aware of how their daily actions affect perception. They must take control of their personal image as well because markets and consumers today want to know the various "faces" of the organization.

Employees and doers at all levels can no longer "hide" behind the company facade, and people expect them to be front and center.

The obvious aspects of organizational image building can't be ignored either. Product and service quality must be maintained and improved to ensure at least a stable if not growing market share. Physical assets, buildings, and grounds must be kept immaculate. The entire customer experience must be regularly evaluated and updated to reflect changes in customer needs and wants.

Over the years, I've been asked a number of times how to creatively develop and maintain a competitive advantage in an environment marked by rapid changes in technology, fluid delivery systems, intense competition, real-time communication, and instant (and often brutal) customer "experience" reports through social media sites. A day doesn't go by that I don't hear about someone bemoaning a poor customer service experience. In fact, I believe customer service has gotten so bad that some people generally seem to expect a bad experience. As a result, I believe we have lowered our bar to the point where we now just tolerate being treated poorly.

The Not-So-Secret, Secret

Although I do think it's becoming more difficult to maintain an advantage, I believe there is a solution…perhaps even the solution. Here's the "secret" competitive advantage solution, especially if you're trying to build a positive personal or organizational image: **Create a culture of _above-and-beyond_ service. This will immediately place you ahead of most, if not all, of the competition.**

Success comes by helping others get what they want. Go beyond the Golden Rule. In other words, treat people BETTER than you would want to be treated.

As a child, my father regularly told me that if you borrow something from someone, always return it in better condition than when you initially received it. By doing this, others will be willing to help you again if you need it. When people come to you for what you provide, they are investing their time and possibly their money. Give them back something of greater value. If people believe you truly care and will take care of them first and foremost, they will follow your leadership, believe in your ideas, and give you their business, even if what you're offering doesn't have all of the latest bells and whistles. **Bottom line:** Regardless of anything else, people's *perception* of an experience still comes down to how they feel.

Some time ago, my wife and I took a short vacation to Telluride, Colorado. We went there to relax, do a little hiking, and enjoy the beautiful scenery (which was breathtaking). Without any prior experience, we booked a room at the Inn at Lost Creek in the adjacent town of Mountain Village. We chose this property *only* because they were pet-friendly, and we had our dog in tow. We knew nothing else about it.

From the moment we arrived, the staff was "over-the-top" friendly. They asked us for our names (including the dog's), and NEVER forgot them. In fact, every time we walked by, they would say "hello" and address us by name. Whenever I walked the dog, they treated her like she was their own, also addressed her by name, and offered her treats (they even provided us a special pet basket at check-in full of treats, a mini-flashlight for night walking, and waste bags).

A member of the staff provided us with a short tour of the property to ensure we knew where all of the amenities were located. He made sure we also knew that all snacks and bottled water were "free" and emphasized that if we ever needed more or anything else to let him know. The complimentary breakfast was incredible, and large, fresh-baked cookies were always available in the main lobby (something I overindulged on because frankly, they were awesome).

Every time anyone saw us over the course of our three days there, we were asked how our stay was going and if we needed anything. It didn't matter who was working at the time. Everyone had the same exceptionally *positive attitude* and treated us with the same high level of care and respect. They even made sure we had bottled water, snacks, and a package of dog treats for the road when we departed.

The staff made us *feel* extremely special—almost as if we were the only guests they had. We were so taken aback by the experience, we found ourselves talking about it the whole time we were there and even after we had left. It was incredible, and because of this level of service, this is now the ONLY place we will ever stay WHEN we return. And, of course, we will now tell everyone we know about the Inn at Lost Creek.

Over the years, I have developed something of an obsession with Mont Blanc pens and products. Although there are many other excellent pens and leather products out there (even some with better features), I have become loyal not only to Mont Blanc, but also to one particular boutique. I've purchased merchandise at a number of different Mont Blanc boutiques across the country, but only the store in Oak Brook, Illinois, truly stands out (this store has sadly since been consolidated into the downtown Chicago store).

When I visited this particular store, not only did my service representative there take care of my wants (this really doesn't classify as a needs purchase), she went way above and beyond EVERY time. Each time I made a purchase, she sent me a very nice, handwritten note in beautiful calligraphy thanking me for my purchase. She sent me cards on special days like birthdays and Christmas, and often for no particular reason but to say she hoped all was going well, all in the same striking penmanship. She quickly resolved any purchase issues and often tossed in free "extras," such as pen and pad refills. She took the time to make me feel special and show that my commitment of money was worth it.

No other person at any other retail store (Mont Blanc or otherwise) had invested this much time into my personal satisfaction. As a result, she had my permanent loyalty. She obviously also cared about her *personal* image, and after receiving the news that she was going to be "downsized," she even took the time to inform me of the closing and personally connected me with my new Chicago-based representative. Now that's service and personal leadership. Given the chance, she is the kind of doer I would hire in an instant.

People value most how you make them feel. They will ultimately act based on those feelings, so give them something worth their investment of support, time, and money. If you make them feel special, they will reward you with their long-term loyalty. By proactively building an image of service to others, you will create a sustainable competitive advantage, despite what others—including your competition—are doing or whatever other label you may have been given.

I guess the milk delivery driver was wrong. What's on the outside needs to match what's on the inside. Cows aren't always cows after all.

Taking a Few More Steps:

• Aspire to be better—not the best. The "best" comes and goes, often replaced by another. If you are trying to simply improve a little each time regardless of what you are doing, the rest will come.

• Support others. They will return the favor.

• Have you shared your goals with those closest to you? If not, do it. How passionate are you about accomplishing them? You will need buy-in from those around you to create a supportive environment, which will enhance your chances of success. Success begets motivation which begets more success.

• Ask your "sounding board" for an honest assessment of how others perceive you. Is their feedback in line with how you want to be viewed? If not, perhaps it's time for a change.

• People naturally deviate toward that which is consistent and reliable. Be consistent. Be reliable.

Last, But Not Least...

Figure 10-1. A plaque commemorating the Apollo 1 fire and loss of crew. Photo taken in 2005.

The first time I visited Kennedy Space Center at Cape Canaveral, Florida, I was awestruck. Standing on the site of Launch Complex 34, the location of both the Apollo 1 fire and the launch of Apollo 7 (the first manned Apollo flight), I vividly remember feeling conflicted. I felt somber because I knew the first Apollo crew had perished on this spot, but also energized as I was surrounded by actual artifacts from some of the greatest imagineering ever realized. Launch Complex 34 represented how incredible problem-solving can occur despite the attempt of purple people eaters to disrupt it.

Like my feelings while at the Cape, this chapter is also conflicted. Individual potential is only limited by the barriers one allows into their life (see Figure 10-2 below). As a completely mental process, successful imagineering requires a mind free from distraction and the barriers created by the many purple people eaters we deal with on a daily basis. It also requires a strategy or method to ensure greater success throughout the thinking process. Both will be addressed here, beginning with some of the inhibitors to effective problem-solving.

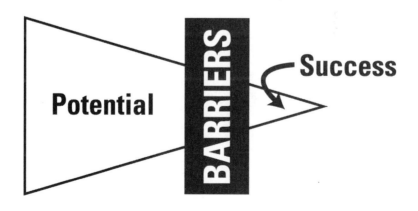

Figure 10-2.

The Good, the Bad, and the Ugly

As a kid, I used to love reading *Mad Magazine*, especially the "Spy vs. Spy" comic, where one spy was always trying to outdo the other. The characters were identical in appearance, other than one was all white in color while the other was all black. Perhaps an insinuation of good versus evil? Not sure. But this concept reflects how I view technology today—two sides of the same thing, nearly always in competition with each other.

Technology has transformed our lives, often in positive and previously unimagined ways. It has the power to enhance business activity, facilitate financial transactions, expedite communication, provide exceptionally realistic visuals (while it entertains), and transport us to distant worlds. However, if left unchecked, technology also possesses the potential to negatively affect how we live and interact as human beings.

Unfortunately, this "bad side" already seems to have engulfed younger generations and is rapidly spreading into older generations, including my own. We are now witnessing the impact on a variety of established social norms and customs, and behaviors once considered unacceptable now seem to be fairly common. This change in behavior can also have a negative impact on imagineering.

For example, consider the following:

• **Too much public disclosure.** Regardless of its direct impact on one's personal identity, the Internet has opened the door to decreased privacy. For instance, most of us have likely witnessed people who seem to have no problem carrying on very personal conversations through texting, email, tweeting, and social media. It's easy to judge others and assume "I would never do that," but have you ever sent a personal email, picture, cartoon, or joke to someone that, if it appeared with your name on the front page of the newspaper, would mortify you? Also, with ever-improving cameras on smartphones, your actions are always under scrutiny (thank you, Moore's Law). During a recent trip to Las Vegas, I saw a t-shirt with the saying, "What happens in Vegas, stays on YouTube." Like it or not, what you once believed to be private is now probably sitting on a server somewhere. Once something enters the Internet, it tends to stay there, and we may have little choice regarding who views it.

• **Changing "rules" for social manners.** Situations once deemed quiet, such as face-to-face conversations, meetings, interviews, classes, weddings, and even funerals are now routinely interrupted by the indiscriminate usage of cell phones. A growing number of people seem more accepting of the practice. It's even worse when people answer their phones regardless of circumstance. Nearly constant emailing and texting to others during actual face-to-face interactions has also become much more common.

• **Reduced face-to-face interaction.** The ease with which technology allows us to communicate can also create an excess of digital content. Sometimes the message itself gets lost in a sea of messages; the back-and-forth of an email or text thread can actually waste more time than making a simple phone call. This practice opens the door to confusion and misunderstanding due to the lack of proper context and verbal/non-verbal cues, provides a tempting platform for people to say nearly anything without having to face the recipient (thus hiding behind their screens), and as a lawyer once told me, creates "evidence." I once spent ten minutes watching my then teenage daughter and four of her friends sitting in my living room texting and not saying a single word to each other. Texting is easy and immediate, but like with most things, there's a price for the convenience.

• **An inability to disconnect.** We love our smartphones and Bluetooth devices. But, when did they become extra appendages we can't seem to live without? They have certainly changed the workplace, often in positive ways, but they have blurred the lines between work and personal life.

• **A condensed life.** Through text messaging, Twitter, and Facebook posts, people have become accustomed to communicating in short, abbreviated statements, which leads to the sharing of immediate, random thoughts (i.e., "theater popcorn—yum!") and "checking-in" when they visit places thus making everyone in their respective network aware of their current location. When daily life is condensed into short bursts of text and tiny amounts of time, the result can be decreased attention spans. Research has shown that young, developing brains can become more habituated than adult brains to constantly switching tasks and less able to sustain attention. According to Michael Rich, an Associate Professor at Harvard Medical School and Executive Director of the Center on Media and Child Health in Boston, "Their brains are rewarded, not for staying on task, but for jumping to the next thing."[1] Patricia Greenfield, a professor at UCLA, looked at more than 50 studies of technology's effects on children and found that technology appears to be damaging critical reasoning and reducing attention span, thus leaving children less skilled at concentrating on a particular point for very long.[2]

• **Poor writing skills.** The new form of abbreviated, acronym-laden writing has directly affected individual writing skills. We've all heard people in business, education, or the media—maybe we've said it ourselves—bemoan the lack of writing skills in people today, and ample evidence can be seen all around us. This lack of skills isn't just reflected in email or other forms of online communication. Like many professors, I've received countless term papers riddled with texting acronyms instead of complete thoughts.

• **The costs of immediate (24/7) feedback.** The results of many decisions can take time to play out, and sometimes a "mental processing" or "cooling off" period is the best thing for solving a conflict. But, with continuous updates, daily opinion polls, unlimited information on the fly, and constant connection, patience doesn't seem to be the virtue it once was. This new form of technological immediacy can lead to the rapid spread of "bad" information, reviews, and growing discontent. It's tough enough to do the hard things and make difficult decisions that require time in order to achieve desired outcomes—and even tougher when people expect "instant."

• **Difficulty differentiating "real" versus "unreal."** Most of us have probably caught ourselves being intrigued by the lifelike qualities and stupendous special effects of video games, even if we don't play and are simply wandering through Target. Yet, is it any wonder that many are increasingly concerned—with good reason—about the apparent decrease in the imagination and creativity of children? Video game sales now surpass those of box office franchises such as *Star Wars* and *Harry Potter*.[3] The biggest difference, however, may be the shift in how we spend our time. An occasional two-hour movie seems like a small chunk of time versus the hours, maybe even days, that some of our kids spend gaming or doing the digital "dead walk" trying to find Pokemon (super popular...for a total of four months).

- **Identity Theft.** These are two words that were never heard together in normal conversation until the advent of the personal computer. Enough said.

I could go on, but I'm sure you get the point. With the huge amount of positives that come with technology, there are an equal number of negatives. We need to be proactively aware of these negatives and how they may ultimately affect our lives, others around us, and our efforts to imagineer. More computing power exists today in a simple musical birthday card than in the entire world 65 years ago. If Moore's Law continues to hold true, and computing power/capacity continues to double every 24 months (or fewer), one can only imagine what wonders or possible problems might be in store for us.

I've had to learn through personal experience what works for me in terms of using technology and what doesn't. Life is complicated enough, and it's not smart or safe to assume technology will always make it easier. I've learned not to automatically accept all new technologies that come along and to proactively assess whether they will help me become more productive and effective. We learn by trying, but effective personal leadership dictates that you must be willing to evaluate something on your own and not allow yourself to be drawn into it because everyone else is using it. In his book, *I'm Working On That*, William Shatner, who played the original Captain Kirk in *Star Trek*, writes:

> And that's kind of the way it happens with technology, isn't it? It's seductive and embraces us until we get used to it and can't do without it...
>
> At first it's a luxury, the domain of early adopters, and then after a while we need this stuff. In fact the reason we get so

*upset when computers don't work is because we've become
so reliant on them. If we weren't reliant, we wouldn't give a
damn. We wouldn't even know they were broken.*[4]

A Totally Different Kind of Distraction

Most parents can probably relate to this scenario (teens too): The parent tries to share with the teen something of value that's been learned through years of personal experience (let's call it wisdom). The teen responds with an eye roll of disregard, some kind of utterance like "whatever," or simply ignores the parent. The parent becomes a little frustrated and presses harder. The teen presses back. Some kind of "discussion" ensues that escalates into something else. The word "disrespect" is either thought or spoken. Sharing time is over (although it may have never really started).

Most parents I know would love to be able to pass on all their collective wisdom to children who would not only listen, but also take this wisdom to heart and keep from repeating the same mistakes. I wasn't one of those of kids. I may have even made life a little more difficult for my parents than most, but I am now one of those parents who, ironically, wants his kids to listen. Although some of what I have written has a parental overtone (from the experience of many personal failures combined with the insights learned from those much wiser), if there is one lesson I wish people would take to heart, including my own children, it would be one related to the use of money.

Money 101

It may seem odd to be discussing money in the context of effective imagineering, but a vivid imagination and effective, creative thinking do not occur in some kind of mental isolation away from the rest of your life. Money can quickly become a huge distraction and driver of increased stress, neither of which allow for a clear mind, free to solve problems, and develop new ideas. Financial problems are HUGE purple people eaters.

People have struggled with this topic since the beginning of documented economic history, and a quick perusal today of the shelves at any bookstore or an online search will yield countless books on the topic. Whether it's a snowball strategy to eliminate existing debt, an investment scheme to maximize returns, or a method to enhance income by selling extra stuff on eBay, there are endless ideas put forth by a great many people.

My insight into the area of personal finance comes directly as a result of personal experience. In fact, I've come to believe I now hold the ultimate secret to financial prosperity, not to mention a mind free from financial stress to work at its highest level—a philosophy that required a large number of financial missteps and much of my life to finally figure out.

Before I share this little gem of insight, I must confess that as a marketing-minded individual, this insight tends to run contrary to the very nature of my efforts to position and promote products to the masses. It also runs contrary to what has become considered "normal" behavior by a majority of people, companies, and our government. It's a lesson I had to learn both through personal experience and the experiences of those close to me. It came to me one day as a life-changing epiphany—a realization that my current financial choices were unsustainable and something had to change. I believe personal financial success can be summed up in this one, simple pearl of wisdom: *Spend less than you make.*

Yes, we've heard this phrase before, but how many of us actually put it into practice? I'm convinced that for most of us (as individuals and as corporate and government entities), sustainable long-term financial success rests with this basic principle. I also know this practice may be a lot easier to say than to actually do.

For many, this philosophy requires a total life paradigm shift—one that includes a radical commitment to changing bad spending habits and making better long- and short-term decisions related to the use of limited resources. Many of the ideas suggested by others will help with this process, but none will ultimately be successful until you start spending less than you make. I suppose an alternative strategy would be to raise your income to a level greater than your spending, but for most of us, reducing the amount we spend is probably easier, especially given today's economic reality.

Everyone I know who has changed spending behavior in this way has, over time, enjoyed reduced and/or no debt, a growth in savings, less stress, and most importantly greater flexibility in decision-making and imagineering. Changing your daily, incremental spending behaviors with the intent of spending less will, over time, help you become more self-sufficient and free to do the things you "want" versus what you "have to do."

Adam Carroll, President of National Financial Educators, author of *The Money Savvy Student,* and coauthor of *Winning the Money Game*, frequently presents what he calls the Four Legacies we must leave future generations. His philosophy begins with creating Financial Freedom, which leads to Time Freedom, because people with money generally have more time. Time Freedom leads to Relationship Freedom, or the ability to spend quality time with those you wish. Finally, Relationship Freedom leads to Service Freedom, or the ability to be of service to others in a way that gives you purpose.[5]

Notice that this entire process begins with creating Financial Freedom. The freedom created by having control of your daily financial choices must start with making the decision to spend less than you make. Many people do not have financial freedom and struggle with realizing the other levels of freedom. Instead they have developed just the opposite—financial stress.

Money is a leading, growing source of stress for Americans. More than three out of every four American families are now in debt today, according to the Federal Reserve's Survey of Consumer Finances.[6] According to a study by the American Psychological Association, three-fourths of the respondents cited money as a significant source of stress in their lives. The Stress in America survey also points to a looming national health crisis caused by the stress created as a result of money-related (76 percent) and work-related (70 percent) issues.[7]

Debt creates financial bondage; this bondage creates stress; and this stress makes effective problem-solving and imagineering extremely difficult. The only real way to reduce debt and the resulting stress is to make the decision to regularly spend less than you make. Remember, most goals are achieved daily, not in a single day. Your debt, like nearly everything else in your life, was more than likely accumulated over time, so this philosophy requires a permanent change in daily spending behavior—a change that will ultimately free your mind to achieve higher levels of creativity.

Pushing Through It

Imagineering new ideas is hard enough without distractions or negative influences, and the result is often increased frustration (think writer's block for an author). Sometimes, a specific strategy or method is required to ensure greater success throughout the thinking process. Many techniques exist today to improve imagineering, and although not an exhaustive list, a few are detailed below:

Brainstorming: What list of imagineering techniques doesn't begin with this popular and perennial favorite? Developed by Alex Osborn in the early 1940s, brainstorming is one of the oldest techniques developed as a way to get a group of minds focused on a single, specific problem to generate a large number of ideas, which could later be evaluated and judged. In his 1948 book, *Your Creative Power*, Osborn listed the following rules for brainstorming:

- Judgment is ruled out.

- "Wildness" is welcome. (Free-wheeling and outrageous)

- Quantity versus quality is wanted. (Write down every idea as all ideas and people are valuable)

- Combination and improvement of ideas are sought.

These same rules still guide brainstorming sessions today. Osborn believed that ideas generated later in the response sequence would typically be more creative. He saw the process of creativity as the unrestricted generation of ideas, which are only later judged as to their "goodness."

However, having been part of multiple brainstorming sessions over the years, I've come to believe the normal approach to the process is somewhat inefficient. In most sessions, the early ideas given out tend to be the most obvious. Once the obvious ideas have been exhausted, people will frequently sit and stare at the ceiling while they think. This time is wasted because members come in "cold," having given little to no prior thought to the problem at hand. If members are provided the problem that needs solved prior to the scheduled session (if possible), I believe the flow of ideas will not only be more efficient, but the ideas themselves will be better as a result.

If you are facilitating a brainstorming session, it's critical to build energy and enthusiasm amongst members while keeping the attention off of yourself. Make the environment playful as this will assist in a greater number of unexpected connections and keep the environment totally "free" from the fear of taking creative risk. While it's important to keep the focus on the problem to be solved, it's also important as the facilitator to pay attention to your feelings concerning the direction and emotional climate of the group. It may take some courage to explore what is *not* being said.[8]

Brainwriting: Just as writing, pen or pencil on paper, slows down your brain and actively engages your senses (see Chapter 7), brainwriting does the same when it comes to making connections. Brainwriting is basically brainstorming on paper and is a continuous, quantity-building process that can be done both individually and in groups. When brainwriting, a list is generated of virtually anything that comes to mind related to the problem to solve. By repeatedly reading through the list, new connections are made, which are then added to the list. As stated earlier, since writing generates more meaning in the mind, it will visually connect your thoughts and feelings like an artist putting paint on a canvas.

When brainwriting in groups, have the participants spend a few minutes writing down their ideas, and then have them pass those ideas on to someone else who will read the list and add to it. This process gets repeated with sheets being exchanged each round between different people. After a determined amount of time, sheets are collected and discussed as a group.

Mindmapping: Mindmapping is a visual technique that uses pen or pencil on paper like brainwriting, but shows immediate connections. A mindmap begins with placing the central topic or problem to be solved in the center of the paper. Through free association, you begin

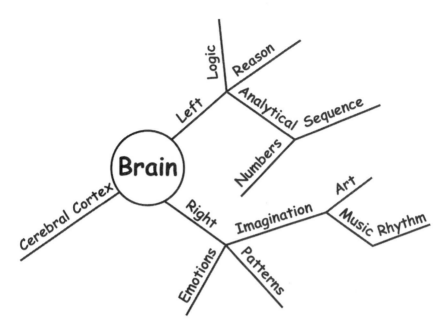

Figure 10-3. A very simple mindmap for the brain.

to branch out from the topic or problem with subtopics, related problems, solutions, or anything else that comes to mind. The same is
done with each of those, and the process continues until any and all
ideas have been exhausted (see the *very* basic mindmap in Figure
10-3 for the topic of the "brain").

If you're a visual thinker, mindmapping will help you get to
the core essence of a topic, problem, or idea. All of the connecting
branches (like a tree) will create a mental framework for organizing
it into a connected whole to see the big picture related to it. To make
the process even more visual, use different-colored pens along with
symbols, illustrations, and pictures.

Forced Relationships: This technique is more of a mental warm-
up prior to actually tackling a real problem. Forced relationships
is a method that looks for common bonds between two or more

previously unconnected things or ideas, like how the Federal Reserve's process for clearing checks was connected to the concept of overnight shipping and how the idea of jet propulsion came from studying the functions of a squid. The juxtaposition of ideas or concepts not normally related is the very essence of creative thinking. By selecting two random items followed by "forcing" them together in a variety of ways, your mind is "warmed up" to the process of imagineering solutions to real problems.

Verbalization: Discovery through verbalization is similar to brainwriting except that it's verbal free association recorded onto your smartphone. With this technique, you talk for a period of time about everything and anything that comes to mind. Tell stories. Describe an object in the room and how you might improve it. Discuss how work might be different if you were in charge. Rant about a problem you need to solve and your feelings surrounding it. The key is to just keep talking. As human beings, we will often say things we later have no recollection of ever saying. In many cases, a creative solution will be "hidden" in the dialogue with yourself. After the recording process, go back, listen, and document.

Documenting Dreams: Becoming consciously aware of your subconscious dreams is often a powerful tool and source of genius. As discussed in Chapter 5, your subconscious works 24/7 and frequently drives those "aha" experiences we have when solving problems. Sometimes, these aha moments are metaphorically embedded into our dreams. While this technique can take time to develop, it has proven itself to be an invaluable tool for some of the world's most creative thinkers. The idea for Dr. Jekyll and Mr. Hyde came to Robert Louis Stevenson in a dream, and Russian chemist Dmitri Mendeleev saw the Periodic Table of Elements in a dream.

Using this approach requires "training" your conscious mind to become more aware of your subconscious mind (similar to how you might wake up in the morning each day prior to your alarm going off because you have trained your conscious mind to be aware of your subconscious internal clock). Each time you awaken (whether it's first thing in the morning or the middle of the night), immediately write down everything—every item, story, fragment, or bit of a dream you can remember. Doing this on a regular basis will gradually improve your ability to consciously recall specific details of all dreams, some of which may prove to be solutions to problems your conscious mind stopped working on some time ago.

While there are many more specific strategies to improve imagineering, I believe a number of normal day-to-day behavioral changes can also greatly assist creative thinking such as:

- **Turn things into a game:** Reframing problems into puzzles or games allows for an open frame of mind and encourages creativity. Fresh ideas often present themselves when you're having fun and feeling relaxed.

- **Challenge the rules:** Habit is a huge restrictor to creative thinking. Since there are always many solutions to all problems, getting out of your routine, breaking your own established rules and patterns, and trying something new can empty your mind of the "right" way of looking at a problem.

- **Allow failure:** Most of us were taught early in life that failure is bad so we have spent most of our lives trying to avoid it. Fear of failure undermines the imagineering process as truly creative ideas often come with some degree of risk. We need to stop taking ourselves so seriously and be open to looking foolish at times, joyously admitting mistakes, and becoming nonconformists.

- **Expect to be creative:** If you think you're not very creative, then you won't be very creative. Think of all the negative feelings we have each day or all the excuses we come up with to not do something. Develop a proactive, positive way of thinking that avoids the excuses and expects good things to happen each day.

- **Keep a journal:** Sometimes an idea you have today has no identified purpose...yet. By keeping track of and writing down your thoughts, ideas, dreams, and anything else of value (good or bad) that occurs in your life each day, you engage your mind through regular writing and store those things for later when they may become more relevant. The same holds true for ideas represented by others found in journal, newspaper, or magazine articles. If you see something that sparks an interest in some way, clip it and stash it in your journal. I frequently go back though my notes and articles stored over the years and find many nuggets of inspiration that now have a purpose.

What's in Your Dash?

I'm a huge believer that effective organizational leadership begins with effective personal leadership. The choices, decisions, and consequences in each context are not mutually exclusive, but mutually connected. Leadership occurs when the behavior of someone is influenced with the intent of achieving something else in an ethical way.

The first "someone" in the sequence must be you. We must be able to make creative decisions and visualize the path toward success; remain focused; allow the necessary time for decisions to play out; possess the work ethic, motivation, and discipline to do what it takes; and assume responsibility for both our physical and digital selves in order to successfully lead others (as well as establish the credibility necessary to gain their confidence).

Most of what I've written in this book is focused almost exclusively on process and personal behavior. I believe true leadership is based on 20% knowledge and 80% action. In other words, successful leadership is far less about what you know and far more about what you do (or sometimes don't do).

Consider a tombstone—a monument to one's life. It usually contains the name and "life years" of someone no longer living. Although the inscription typically focuses on the years when a person was born and subsequently passed away, a person's life is actually represented by the "dash" in between (i.e., 1964–2042).

This dash represents the essence of our lives—the succession of joys, sorrows, successes, failures, and the multitude of other life experiences. Although tombstones don't tell the full story, the "dash" is where life actually happens. The life of the Apollo Program was from 1961–1975. The dash of that life included a quarter-million step journey to the Moon made an amazing nine times.

If you could write the story of your dash, how would it read? Would it be full of regrets for the things you did or didn't do? Or would it be a tribute to all that you attempted to do, be, and accomplish while you were alive?

The choice is yours. Life is framed by a start and finish. Whether or not you are living your desired dash or just dashing your way through life comes down to the choices made within your limited time on Earth. Many people have good intentions but aren't necessarily good with actions. They often act as if they have all the time in the world left to create the story of their dash.

What if you knew exactly when your time was up? Would you live life any differently? Would you make different choices? The decision to take control, be a proactive personal leader, and make the necessary transformative change is ultimately yours.

So make it. The rest of your life begins today.

Taking a Few More Steps:

• Over the next few weeks (whenever you have a "free" moment), select two completely random objects around you and attempt to force connections between them (like trying to jam a square peg in a round hole). Don't judge the quality of the ideas; just have fun with it and bring out your inner "MacGyver."

• Think about your daily routine. In what ways can you challenge the rules, do something different, allow failure, and expect to be more creative in your life?

• Document (journal) all of your ideas on a continuous basis whether you "feel" they have any value or not. You never know when one of them could become a game changer for you.

• Ask yourself: "What do I really want? When I reach the end of my life, will I be glad that I got it or did it?"

• List five people you believe have "great lives." Why do you think so? What do they have in common? How are they different from yours?

Acknowledgements

First, I would like to acknowledge my wife, **Robin**…my best friend, honest critic, huge supporter, and biggest fan.

I'd like to acknowledge my parents, **Don** and **Sandy Paustian**. Not only did their youthful exuberance bring me into the world, they also provided me the foundation and values that serve as the basis for this book (not to mention a bunch of the stories).

Special thanks to Apollo 15 astronaut, **Alfred Worden**, for being an inspiration during my youth and for writing the Foreword for this book.

Without the help of my creative team, this book wouldn't have been worth its price in paper. Thanks to my content editor, **Sara Stibitz**, for helping to wordsmith my ideas, my brother, **Michael Paustian**, for his eye-catching and spot-on cover design, and my proof editors, **Maggie** and **David Wiant** and **Dr. Kim Linduska** for catching those "unseen" typos and errors throughout the final manuscript.

I'd like to thank my mastermind team for their thoughts, edits, and support throughout the process. This group includes **Danny Beyer**, **Tom Henricksen**, **Matt McKinney**, **Ted Nahas**, and **Randy Tennison**.

Finally, I'd like to express my gratitude to those who read the book and provided quotes and feedback during the its development: **Eugene Cernan** (Gemini 9, Apollo 10/17 astronaut and the last man to walk on the Moon), **Alan Bean** (Apollo 12, Skylab astronaut and the fourth man to walk on the Moon), **Alfred Worden** (Apollo 15 astronaut and aeronautical engineer), **Dr. David Gallo** (oceanographer and mission leader to the wreckage of the RMS Titanic), **Howard Berger** (Oscar and Emmy award-winning make-up and effects designer), **Kari Byron** (former co-host of Mythbusters and co-host of the White Rabbit Project on Netflix), and **Bill Penczak** (businessman and marketing guru).

References

CHAPTER 1

1. *The How the Apollo Program Produced Economic Wealth.* Retrieved September 15, 2016, from the EIR Science & Technology website: http://www.larouchepub. com/eiw/public/1987/eirv14n21-19870522/eirv14n21-19870522_024- how_the_apollo_program_produced.pdf

2. Lyons, Lauren (2013, July 10) *5 Popular Misconceptions About NASA.* Retrieved September 15, 2016, from the Wrike.com website: http://www.huffing- tonpost.com/lauren-lyons/misconceptions-nasa_b_3561205.html

3. Bonnie, Emily (2015, July 1) *9 Project Management Lessons Learned from the Apollo 11 Moon Landing.* Retrieved September 15, 2016, from the Huffington Post website: https://www.wrike.com/blog/project-management-lessons-learned- apollo-11-moon-landing

4. *The Continued Socioeconomic Impact of Apollo 11.* Retrieved September 15, 2016, from the Online Learning Tips website: http://onlinelearningtips.com/ 2014/07/25/the-continued-socioeconomic-impact-of-apollo-11

CHAPTER 2

1. *NASA - Excerpt from the 'Special Message to the Congress on Urgent National Needs'*. Retrieved June 5, 2016, from the NASA website: https://www.nasa.gov/vision/space/features/jfk_speech_text.html#.V1WOq-mv38s

2. *Project Apollo: A Retrospective Analysis*. Retrieved June 5, 2016, from the NASA website: http://history.nasa.gov/Apollomon/Apollo.html

3. Zagursky, E. (2010, July 14). *Professor Discusses America's Creativity Crisis in Newsweek*. Retrieved November 18, 2010, from the College of William & Mary website: http://www.wm.edu/news/stories/2010/professor-discusses-americas-creativity-crisis-in-newsweek-123.php

4. (2010, May 18). *IBM 2010 Global CEO Study: Creativity Selected as Most Crucial Factor for Future Success*. Retrieved November 18, 2010, from the IBM website: http://www.03.ibm.com/press/us/en/pressrelease/31670.wss

5. *Inside Views: A Look at the Huge Upswing in China Patent Filings*. Retrieved June 5, 2016, from the Intellectual Property Watch website: http://www.ip-watch.org/2015/04/22/a-look-at-the-huge-upswing-in-china-patent-filings

6. *China Emerges as World Patent Leader*. Retrieved June 5, 2016, from the Thomson Reuters website: http://thomsonreuters.com/en/articles/2014/china-emerges-as-world-patent-leader.html

7. Paustian, Anthony. (1997). *Imagine! Enhancing Your Critical Thinking and Problem-Solving Skills*. Upper Saddle River, NJ: Prentice Hall.

8. *Francium*. Retrieved January 25, 2011, from the Elementymology & Elements Multidict website: http://elements.vanderkrogt.net/element.php?sym=fr

9. (2008, October 9) *Fred Smith: An Overnight Success*. Retrieved November 9, 2014, from the Entrepreneur website: http://www.entrepreneur.com/article/197542

10. Brown, Abram (2014, January 23) *10 Things You Might Not Know About FedEx Billionaire Fred Smith*. Retrieved November 9, 2014, from the Forbes website: http://www.forbes.com/sites/abrambrown/2014/01/23/10-things-you-might-not-know-about-fedex-billionaire-fred-smith

11. *Samuel Colt*. Retrieved June 10, 2016, from the History Channel website: http://www.history.com/topics/inventions/samuel-colt

12. Bellis, M. *The History of Commercial Deodorants*. Retrieved January 28, 2011, from the About.Com Inventors website: http://inventors.about.com/od/dstartin-ventions/a/deodorants.htm

13. *Thomas Edison*. Retrieved June 10, 2016, from the History Channel website: http://www.history.com/topics/inventions/thomas-edison

14. (1996). *Thomas Edison Quotes*. Retrieved February 6, 2011, from the Thomas Edison.com website: http://www.thomasedison.com/quotes.html

15. *Imagineering*. Retrieved June 10, 2016, from the Wikipedia website: https://en.wikipedia.org/wiki/Imagineering

16. (2007, December 26). *Invention of the Telephone Switch*. Retrieved January 29, 2011, from the Strowger website: http://www.strowger.com/about-us/strowger-invention-of-telephone-switch.html

17. *John Dunlop*. Retrieved January 29, 2011, from the Great Idea Finder website: http://www.ideafinder.com/history/inventors/dunlop.htm

18. Suddath, C. (2009, June 23). *A Brief History of Kodachrome*. Retrieved January 29, 2011, from the Time website: http://www.time.com/time/arts/article/0,8599,1906503,00.html

19. *Female Inventors–Hedy Lamarr*. Retrieved January 29, 2011, from the Inventors website: http://www.inventions.org/culture/female/lamarr.html

20. *Edison's Miracle of Light*. Retrieved January 29, 2011, from the Public Broadcasting Service (PBS) website: http://www.pbs.org/wgbh/amex/edison/film-more/description.html

21. *The Quartz Watch*. Retrieved February 6, 2011, from the Smithsonian website: http://invention.smithsonian.org/centerpieces/quartz/inventors/index.html

22. Tang, K. (2010, September 28). *Flat Panel TV History*. Retrieved February 6, 2011, from the eHow website: http://www.ehow.com/facts_7250075_flat-panel-tv-history.html

23. *The Apollo Program*. Retrieved May 15, 2010, from the National Aeronautics and Space Administration (NASA) website: http://history.nasa.gov/apollo.html

24. Saran, Cliff. (2009, July). *Apollo 11: The Computers That Put a Man on the Moon*. Retrieved June 10, 2016, from the Computer Weekly website: http://www.computerweekly.com/feature/Apollo-11-The-computers-that-put-man-on-the-moon

25. *World's Fair History*. Retrieved January 27, 2011, from the EXPO Museum website: http://www.expomuseum.com

26. *Iowa View: Will America Again Pursue New Frontiers?* Retrieved August 1, 2016, from the FLORIDA SPACErePORT website: http://spacereport.blogspot.com/2012_08_01_archive.html

27. Cernan, Eugene, Personal Communication, March 5, 2014.

28. Cernan, Eugene, Personal Communication, July 23, 2014.

28. *Tomorrowland: Walt's Vision for Today*. Retrieved June 10, 2016, from the Walt Disney website: http://waltdisney.org/exhibitions/tomorrowland

CHAPTER 3

1. Whitfield, S.E. & Roddenberry G. (1968). *The Making of Star Trek*. New York, NY: Ballantine Books.

2. *List of Star Trek Races*. Retrieved October 14, 2011, from Wikipedia, the Free Encyclopedia website: http://en.wikipedia.org/wiki/List_of_Star_Trek_races

3. Paustian, Terry. Personal Interview, October 9, 2011.

4. *Box Office History for Star Trek Movies*. Retrieved June 10, 2016, from The Numbers website: http://www.the-numbers.com/movies/franchise/Star-Trek#tab=summary

5. *Motion Pictures By Independent Filmakers*. Retrieved October 13, 2011, from the American United Entertainment website: http://www.americanunitedent.com/ifi.htm

6. *The Blair Witch Project (1999) Trivia.* Retrieved October 13, 2011, from the Internet Movie Database (IMDB) website: http://www.imdb.com/title/tt0185937/trivia

7. *In Saturn's Shadow.* Retrieved October 17, 2011, from the NASA Jet Propulsion Laboratory website: http://photojournal.jpl.nasa.gov/catalog/PIA08329

8. *Inventors of the Modern Computer.* Retrieved October 18, 2011, from the About.com Inventors website: http://inventors.about.com/library/weekly/aa121598.htm

9. Jones, D. *Punched Cards.* Retrieved October 2, 2011, from the University of Iowa website: http://www.divms.uiowa.edu/~jones/cards/history.html

10. Moore, G.E. (1965, April 19). *Cramming More Components onto Integrated Circuits.* Retrieved October 18, 2011, from the Intel website: ftp://download.intel.com/museum/Moores_Law/ArticlesPress_Releases/Gordon_Moore_1965_Article.pdf

11. Fries, A. (2010, January 19). *The Dynamic Duo: Imagination + Knowledge.* Retrieved September 29, 2011, from the Psychology Today website: http://www.psychologytoday.com/blog/the-power-daydreaming/201001/the-dynamic-duo-imagination-knowledge

12. *Edwin Land Biography.* Retrieved June 10, 2016, from Biography website: http://www.biography.com/people/edwin-land-9372429

13. *Square Fruit Stuns Japanese Shoppers.* Retrieved June 10, 2016, from BBC News website: http://news.bbc.co.uk/2/hi/asia-pacific/1390088.stm

14. Handel, A. (Producer), & Jones, J. (Director). (2005). *How William Shatner Changed the World.* [Film]. United States: Allumination Filmworks.

15. Chang, Kenneth. *The Long, Strange Trip to Pluto, and How NASA Nearly Missed It.* Retrieved July 12, 2016, from the New York Times website: http://www.nytimes.com/2015/07/19/us/the-long-strange-trip-to-pluto-and-how-nasa-nearly-missed-it.html?_r=0

16. Chang, Kenneth. *NASA's New Horizons Spacecraft Sends Signal from Pluto to Earth.* Retrieved July 12, 2016, from the New York Times website: http://www.nytimes.com/2015/07/15/science/space/nasa-new-horizons-spacecraft-reaches-pluto.html

CHAPTER 4

1. Golf balls in the jar—The philosophy professor. (2011). *Pickchur*. Retrieved from www.pickchur.com/2011/03/ golf-balls-in-the-jar-the-philosophy-professor

2. Active listening: Hear what people are really saying. (2015). *Mind Tools*. Retrieved from www.mindtools. com/CommSkll/ActiveListening.htm

3. Cernan, Eugene, Personal Communication, July 23, 2014.

4. Bean, Alan, Personal Communication, March 4, 2010.

5. Chabris, C. & Simons, D. (2010a). *The gorilla experiment. The Invisible Gorilla*. Retrieved from www.theinvisiblegorilla.com/gorilla_experiment.html

6. Chabris, C. & Simons, D. (2010b). Videos. *The Invisible Gorilla*. Retrieved from www.theinvisiblegorilla.com/videos.html

7. Cernan, Eugene, Personal Communication, July 23, 2014.

CHAPTER 5

1. Soniak, Matt. *How One Dad Got Lawn Darts Banned*. Retrieved February 4, 2015, from the Mental Floss website: http://mentalfloss.com/article/31176/how-one-dad-got-lawn-darts-banned

2. *The Awful Truth About Lawn Darts*. Retrieved February 4, 2015, from the Democratic Underground website: http://www.democraticunderground.com/discuss/duboard.php?az=view_all&address=118x50664

3. *The History of Arm & Hammer*. Retrieved August 24, 2015, from the Arm & Hammer website: http://www.armandhammer.com/aboutus.aspx

4. *The SPAM Story*. Retrieved February 4, 2015, from the SPAM website: http://www.spam.com/spam-101/history-of-spam

5. Grabianowski, Ed. *How Twinkies Work*. Retrieved February 4, 2015, from the How Stuff Works website: http://science.howstuffworks.com/innovation/edible-innovations/twinkie2.htm

6. *Electronic Communications Privacy Act of 1986 (ECPA).* Retrieved June 3, 2016, from the U.S. Department of Justice website: https://it.ojp.gov/privacyliberty/authorities/statutes/1285

7. Dunne, David. (2010, November 10). *United Breaks Guitars: Case Study for the Rotman School of Management, University of Toronto.* Retrieved January 30, 2015, from the Right Side of Right website: http://www.rightsideofright.com/wp-content/uploads/2010/03/United-Breaks-Guitars-Case-Jan-11-10-21.pdf

8. *United Breaks Guitars.* Retrieved January 30, 2015, from the Wikipedia website: http://en.wikipedia.org/wiki/United_Breaks_Guitars

9. Doering, Christopher. (2015, June 24). "The New Dimension of Food." *Des Moines Register.*

10. McMahon, Bucky. (2015, November). "These are the Droids We're Looking for." *GQ Magazine.*

11. Easton, Nina. (2012, January 16). "Fortune's Guide to the Future." *Fortune.*

12. Hoang, Kristine (March 22, 2016). *AAP: Ebook Sales Down 12.7 Percent.* Retrieved October 28, 2015, from the Digital Book World website: http://www.digitalbookworld.com/2016/aap-ebook-sales-down-12-7-percent/

13. Murphy, Mark. *'SMART' Goals Can Sometimes Be Dumb.* (January 8, 2015) Retrieved May 25, 2016, from the Forbes website: http://www.forbes.com/sites/markmurphy/2015/01/08/smart-goals-can-sometimes-be-dumb/ - 131ed902 142c

14. Tynan, Dan. *10 of Tech's Biggest Missed Opportunities.* (August 19, 2009) Retrieved May 25, 2016, from the IT Business website: http://www.itbusiness.ca/news/10-of-techs-biggest-missed-opportunities/13860

15. La Monica, Paul R. *Why Facebook could one day be worth $1 Trillion.* (April 28, 2016) Retrieved May 25, 2016, from the CNN website: http://money.cnn.com/2016/04/28/investing/facebook-trillion-dollar-market-value

16. Tynan, Dan. *10 of Tech's Biggest Missed Opportunities.* (August 19, 2009) Retrieved May 25, 2016, from the IT Business website: http://www.itbusiness.ca/news/10-of-techs-biggest-missed-opportunities/13860

17. Wilford, John Noble (1969). *We Reach the Moon: The New York Times Story of Man's Greatest Adventure.* New York: Bantam Paperbacks.

18. *John F. Kennedy Moon Speech – Rice Stadium.* Retrieved June 4, 2016, from the NASA website: http://er.jsc.nasa.gov/seh/ricetalk.htm

19. *Benefits from Apollo: Giant Leaps in Technology.* Retrieved May 25, 2016, from the NASA website: https://www.nasa.gov/sites/default/files/80660main_ApolloFS.pdf

20. *Leadership IQ Study.* Retrieved May 25, 2016, from the LeadershipIQ website: http://www.leadershipiq.com/blogs/leadershipiq/35353793-are-smart-goals-dumb

CHAPTER 6

1. *Apollo-1 (204).* Retrieved June 20, 2016, from the NASA website: http://history.nasa.gov/Apollo204

2. *13 Things That Saved Apollo 13, Part 10: Duct Tape.* Retrieved June 20, 2016, from the Universe Today website: http://www.universetoday.com/63673/13-things-that-saved-apollo-13-part-10-duct-tape

3. *Dreaming a Different Apollo.* Retrieved June 15, 2016, from the Wired website: http://www.wired.com/2014/10/dreamingadifferentapollo

4. *Falcon 9 & Dragon to Return Astronauts to Space.* Retrieved June 20, 2016, from the SpaceX website: http://www.spacex.com/falcon9

5. *What is Orion?* Retrieved June 20, 2016, from the NASA website: http://www.nasa.gov/audience/forstudents/5-8/features/nasa-knows/what-is-orion-58.html

6. Paustian, Anthony. (1997). *Imagine! Enhancing your critical thinking and problem-solving skills.* Upper Saddle River, NJ: Prentice Hall

7. Heath, Chip and Heath Dan. (2008). *Made to Stick.* New York, NY: Random House, Inc.

8. Marantz, Steve (2002). *History of the World Series – 1997.* Retrieved February 16, 2010, from the Sporting News website: http://www.sportingnews.com/archives/worldseries/1997.html

CHAPTER 7

1. Hamilton, Jon. *Think You're Multitasking? Think Again.* (October 2, 2008) Retrieved October 28, 2015, from the NPR website: http://www.npr.org/templates/story/story.php?storyId=95256794

2. *Is $50,000 Enough to Buy Happiness? What about $161,810?* (April 2013) Retrieved January 11, 2016, from the Fast Company website: http://www.fastcompany.com/3006746/is-50000-enough-to-buy-happiness-what-about-161810

3. *The Dangers of Image Crafting.* Retrieved January 11, 2016, from the Whole9 website: http://whole9life.com/2014/03/dangers-image-crafting

4. Shapiro, A. & Campbell, L. (2014). *The Book of Odds: from Lightning Strikes to Love at First Sight, the Odds of Everyday Life.* New York, NY: William Morrow, Inc.

5. (2015). "Brain 101." Issue 70, *360 Magazine.* Steelcase, Inc.: Grand Rapids, Michigan.

6. Vaughan, Michael. *Know Your Limits, Your Brain Can Only Take So Much.* (January 24, 2014) Retrieved October 28, 2015, from the Entrepreneur website: http://www.entrepreneur.com/article/230925

7. Bradberry, Travis. *How Successful People Make Smart Decisions.* (October 7, 2015) Retrieved October 28, 2015, from the Forbes website: http://www.forbes.com/sites/travisbradberry/2015/10/07/10-tricks-successful-people-use-to-make-smart-decisions

8. *Building the Educated and Employed Communities of Tomorrow.* Retrieved December 28, 2015, from the GED Testing Service website: http://www.gedtestingservice.com/uploads/ files/9bce820a49287fec1febad56e98bccef.pdf

9. *Why Does Writing Make Us Smarter?* (July 16, 2011) Retrieved October 2, 2015, from the Huffington Post website: http://www.huffingtonpost.com/2011/07/16/why-does-writing-make-us-_n_900638.html

10. Cringely, R. (Screenplay), & Gau, J., Segaller, S. (Producers). *Triumph of the Nerds.* [Film]. United States: PBS.

11. Stieff, Quintin. Sermon. Valley Church, West Des Moines. (January 11, 2014). Speech.

12. (2015). "Think Better." Issue 70, *360 Magazine*. Steelcase, Inc.: Grand Rapids, Michigan.

CHAPTER 8

1. *The American Heritage Dictionary of the English Language*, 4th Edition. (2010). Boston, MA: Houghton Mifflin Harcourt Publishing Company.

2. Rhian, Jason (April 15, 2015). *Apollo's Haise and Cunningham: We Got the Glory, But the Workers Got Us to the Moon*. Retrieved June 24, 2016, from the Spaceflight Insider website: http://www.spaceflightinsider.com/space-flight-history/apollos-haise-and-cunningham-we-got-the-glory-but-the-workers-got-us-to-the-moon

3. Worden, Alfred. Personal Communication, September 19, 2016.

4. Maxwell, John. (1998). *The 21 Irrefutable Laws of Leadership*. Nashville, TN: Thomas Nelson Publishers.

5. *Gazette Classics: Kurt Warner File* (Select Articles from Gazette Text Archives, 1998-2008). Retrieved November 4, 2010, from the WordPress website: http://lookinginatiowa.wordpress.com/gazette-classics-kurt-warner-file

6. Belle, A. *Kurt Warner: Biography, Facts, and Stats*. Retrieved November 4, 2010, from the FreeResource website: http://www.thefreeresource.com/kurt-warner-biography-facts-and-stats

7. Warner, K. & Silver, M. (2000). *All Things Possible*. New York, NY: HarperCollins Publishers.

8. Rodgers, Melissa. Personal Communication, July 19, 2010.

9. *James Dyson*. Retrieved November 5, 2010, from the Great Idea Finder website: http://www.ideafinder.com/history/inventors/dyson.htm

10. *Chester F. Carlson*. Retrieved November 5, 2010, from the Great Idea Finder website: http://www.ideafinder.com/history/inventors/carlson.htm

11. *Seussville: Dr. Seuss's Biography.* Retrieved November 5, 2010, from the Seussville website: http://www.seussville.com/#/author

12. Carlin, G. (1997). *Brain Droppings.* New York, NY: Hyperion.

13. Organ, D., Podsakoff, P. & MacKenzie, S. (2006). *Organizational Citizenship Behavior.* Thousand Oaks, CA: Sage Publications, Inc.

14. Zabloski, J. (1996). *The 25 Most Common Problems in Business (and How Jesus Solved Them).* Nashville, TN: Broadman & Holman Publishers.

15. *Babe Ruth.* Retrieved December 3, 2010, from the Baseball-Reference website: http://www.baseball-reference.com/players/r/ruthba01.shtml

16. Rosner, B. (2005, February 25). *Working Wounded: Getting Pink Slipped.* Retrieved December 3, 2010, from the ABC News website: http://abcnews.go.com/Business/WorkingWounded/story?id=547848

17. Beals, G. (1999). *The Biography of Thomas Edison.* Retrieved December 3, 2010, from the Thomas Edison website: http://www.thomasedison.com/biography.html

18. *Michael Jordan Quotes.* Retrieved December 3, 2010, from the Brainy Quote website: http://www.brainyquote.com/quotes/quotes/m/michaeljor127660.html

19. *Lee Iacocca.* Retrieved December 3, 2010, from the Encyclopedia of World Biography website: http://www.notablebiographies.com/Ho-Jo/Iacocca-Lee.html

20. Ziskin, L. (Producer), Williams, B. (Producer), & Oz, F. (Director). (1991). *What About Bob?* [Motion Picture]. United States: Touchstone Pictures. Used with permission.

21. Ecclesiastes 7:8, *Holy Bible, New Century Version.* (2003). Nashville, TN: Thomas Nelson, Inc.

CHAPTER 9

1. *Labor Force Statistics from the Current Population Survey (1969).* Retrieved July 25, 2016, from the United States Department of Labor – Bureau of Labor Statistics website: http://data.bls.gov/pdq/SurveyOutputServlet

2. *The Apollo Program*. Retrieved May 15, 2010, from the National Aeronautics and Space Administration (NASA) website: http://history.nasa.gov/apollo.html

3. Haake, A. (2009, July)."Time Machine – March 1962." *Popular Mechanics*, 21.

4. *The Alamo - History*. Retrieved May 18, 2010, from the Official Alamo website: http://www.thealamo.org/history.html

5. Gruver, E. (1997). "The Lombardi Sweep." *The Coffin Corner*, Vol. 19, No. 5. http://www.profootballresearchers.org/Coffin_Corner/19-05-712.pdf

6. Bean, Alan, Personal Communication, March 4, 2010.

7. Cernan, Eugene, Personal Communication, March 5, 2014.

8. Worden, Alfred, Personal Communication, July 21, 2016.

9. Zickuhr, K. (2010, December 16). *Generations Online in 2010*. Retrieved July 25, 2011, from the Pew Internet Research website: http://pewinternet.org/Reports/2010/Generations-2010/Overview.aspx

10. Protess, B., Rusli, E., & Craig, S. (2011, March 31). "Buffet's Handling of Deputy Baffles Some Experts." Retrieved June 21, 2011, from the *New York Times* website: http://dealbook.nytimes.com/2011/03/31/buffetts-handling-of-deputy-baffles-some-experts/?partner=rss&emc=rss

11. Bissinger, B. (2010, February). "Tiger in the Rough." Retrieved July 27, 2011, from the *Vanity Fair* website: http://www.vanityfair.com/culture/features/2010/02/tiger-woods-201002?printable=true

12. (2010, February 24). "Toyota's Humbling Fall." Retrieved July 27, 2011, from the *CBS News* website: http://www.cbsnews.com/stories/2010/02/24/evening news/main6240051.shtml

13. (2010)." Toyota's Fall is Speeding Out of Control." Retrieved July 27, 2011, from the *Motor Trend* website: http://www.printfriendly.com/print/v2?url= http%3A%2F%2Fblogs.motortrend.com%2Ftoyotas-fall-is-speeding-out-of-control-2592.html

14. Jensen, C. (2010, June 18). "Toyota's Image Falls in J.D. Power Survey," *New York Times*, sec. B, 5.

15. Beyer, Danny (2014). *The Ties That Bind: Networking with Style*. Des Moines, IA: BookPress Publishing.

16. (2011, July 13). *comScore Releases June 2011 U.S. U.S. Search Engine Rankings.* Retrieved July 27, 2011, from the PR Newswire website: http://www.prnewswire.com/news-releases/comscore-releases-june-2011-us-search-engine-rankings-125513223.html

CHAPTER 10

1. Richtel, Matt. (2010, November 21). "Growing Up Digital, Wired for Distraction." Retrieved November 26, 2011, from the *New York Times* website: http://www.nytimes.com/2010/11/21/technology/21brain.html?pagewanted=all

2. Lister, John (2009, January 30). *Technology Has Mixed Effects on Child Development, Research Suggests.* Retrieved November 26, 2011, from the Info Packets website: http://www.infoPackets.com/news/technology/it/2009/20090130_technology_has_mixed_effects_on_child_development_research_suggests.htm

3. Snider, M. (2011, November 18). 'Call of Duty' sells $775M in 5 days, rivals movie giants. *USA Today*, 1A.

4. Shatner, William & Walter, C. (2004). *I'm Working on That*. New York, NY: Pocket Books, Inc.

5. Carroll, Adam, Personal Communication, March 4, 2016.

6. Soong, J. (2010). *The Debt-Stress Connection*. Retrieved November 26, 2011, from the Web MD website: http://www.webmd.com/balance/features/the-debt-stress-connection

7. Hickey, J. (2010, November 10). "Stress Over Money, Work, Economy Top the List for Americans." Retrieved November 26, 2011, from the *ABC News* website: http://www.abcnews.go.com/Business/American-psychological-association-2010-stress-america-survey/story?id=12108276#.TtG9fmBR6CY

8. Paustian, Anthony. (1997). *Imagine! Enhancing Your Critical Thinking and Problem-Solving Skills*. Upper Saddle River, NJ: Prentice Hall.

Brand Acknowledgements

The following brands and corporate names used throughout the book are all registered trademarks of the following:

Alcoa	Harry Potter	NBA	Star Wars
Amazon	Hewlett Packard	New York Yankees	Starbucks
Amp Energy	Hilton	Nike	Subway
Apple	History Channel	Nuclear Energy	Taylor
Arm & Hammer	Houghton Mifflin	Oreck	Twinkies
Chicago Bulls	IBM	Pearson	Twitter
CNN	IMAX	Pokemon	Union Carbide
Disney	Indian Motorcycles	Polaroid	United Airlines
Dyson	Instagram	Pottery Barn	Vanguard Press
Facebook	Kindle	RCA	Venom Energy
FedEx	Kirby	Red Bull Energy	Wired
Flickr	Kodak	Red Robin	Xerox
Ford Motor Co.	LinkedIn	Seiko	Yahoo
Full Throttle	Microsoft	Sharp	YouTube
General Electric	MoMA	Simon & Schuster	
Google	Monster Energy	SPAM	
Hamilton Watch	Motorola	Star Trek	

ATTENTION CORPORATIONS, UNIVERSITIES, COLLEGES, AND PROFESSIONAL ORGANIZATIONS: